What Engineers Should Know About Fibre Composites

by
Michael R. Piggott
A.R.C.S., Ph.D.
SAMPE Fellow (Retired)
Chemical Engineering and Applied Chemistry,
University of Toronto, Canada.

Published by MERP Enhanced Composites Inc.
Toronto, Canada.

ISBN 978-0-9783615-0-1

Frontispiece: A notable application of the first glass-polymer: the 107m tall radome that resides at the top of the Toronto 550m CN Tower. (Courtesy CN Tower Ltd.)

What Engineers Should Know About Fibre Composites

5. Environmental Effects

6. How to Make and How to Use Composites

Appendices

Index

Introducing the author

Preface

This book is based on more than 40 years of working in the field of Advanced Composites, first in Atomic Energy of Canada's Chalk River Laboratories and then at the University of Toronto. The first edition of Load Bearing Fibre Composites (Pergamon, 1980) was based on courses of lectures given to upper-year and graduate students in the engineering departments of the University. Since then the subject has grown up, and whole departments and specialist centres are devoted to the study of it.

Naturally, with all this activity, much more is known, but very rarely have the initial precepts been re-examined. In the words of one President of the American Society for Composites; "we just move on". But when writing my first book, I realized that many of the theoretical concepts worked out in it rested on evidence that was either very slender or non-existent. So an extensive and thoroughgoing experimental program was embarked upon to find new evidence. The results, described in more than 100 papers, are summarized in the Second Edition (Kluwer, 2002, hereinafter referred to as LBFC2), which is more than the twice as long as the first edition.

So this slim book attempts to present only what you need to know. Three aspects of the subject that appear not to be of interest to the practicing engineer have received scant attention herein.

1) The fibre/polymer interface (or interphase). The manufactures of the raw materials appear to have got it all under control and the underlying theory appears to be redundant due to problems with the shear failure processes envisaged (see Section 4.1).

2) Reinforced metals. Because the high temperatures that have to be used to make them, leave them with lethally high internal stresses. Also thermal cycling fatally weakens them.

3) Reinforced ceramics. Unless you choose a fibre with the same composition as the matrix they also have very high internal stresses. But even with fibres of the same composition as the matrix, the brittleness of the matrix makes the material unserviceable for significant tensile loads.

The topic of compression strength, still much misunderstood, has been explained in the light of the latest research results. The deep seated misconceptions about the shear strength (*non-existent*), and the inefficacy of angle ply laminates (*extremely effective*) have been rectified and highlighted by illustrations from recent experimental results.

The earlier books contained problems at the end of each chapter, and many teachers have told me they appreciated them. So I have used some of them here, and contrived others. Solutions are available (to certified teachers) in a separate book.

It only remains for me to thank the people and organizations that freely provided information and pictures, and my colleagues – worldwide – for lending their ears to my sometimes preposterous thoughts, and despite this, giving invaluable advice, and my students whose dedication to the offbeat projects that I gave them to pursue provided the evidence needed to put the subject on a firm theoretical basis. Last but not least, my erstwhile secretary, Jenny Clifford who encouraged my students for so many years, so enhancing their efforts and thus making this book possible.

Michael Piggott, Fall, 2006.

Chapter 1. Mechanics of Isotropic and Anisotropic Materials

In this introductory chapter we set out to do three things: define the basic concepts needed for the development of the reader's understanding of composites; introduce the special terminology and symbols used in the book; and describe some of the test methods used.

1.1. Stresses, strains and strength

Since the book is concerned with composites as materials for use in situations where significant loads are to be carried, we need to quantify the effects of these loads on materials in general. A ten ton weight will squash a small piece of material, but if we make it big enough, the same material will easily withstand the load. So we take the size into account by defining *stress* as *force per unit of area*, as shown in Fig. 1.1. Thus

$$\sigma = F_t / A \tag{1.1}$$

Tensile stresses are positive, while a negative value defines a compressive stress. These are the *axial stresses*.

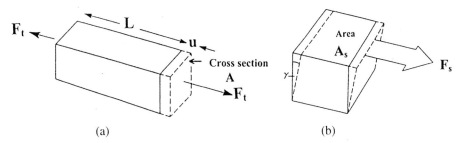

Figure 1.1. (a) A rod in axial tension. (b) Simple shear.

The basic units of stress are Nm^{-2}, or Pascals (Pa), but since this is so small we normally use millions – MPa – or billions – GPa. The symbol used for tensile and compressive stresses is σ, and subscripts are used to indicate the direction of the stress. The cartesian set x,y,z is shown in Fig. 1.2a, but the numbers 1, 2 and 3 are also used, especially for anisotropic materials such as fibre composites.

In this book we will use 1 to indicate the fibre direction in a unidirectional lamina, with 2 and 3 reserved for the other directions, as shown in Fig. 1.2b. In addition, subscripts are used to indicate special values. Thus, for the ultimate tensile strength we use u. For example σ_{1u} indicates the strength of a lamina in the fibre direction. Displacements are u, v and w, respectively in the x, y and z directions, and in the 1, 2 and 3 directions A full symbols list appears in Appendix A where it will be noticed that the subscripts have many other uses.

Shear stress is illustrated in Fig. 1.1b. (A negative value is commonly used to indicate clockwise rotation, but this is not always necessary and will not be used here.)

$$\tau = F_s / A_s \tag{1.2}$$

Again subscripts are used to indicate directions, but this time we need two – see Fig. 1.2a. The first subscript defines the plane on which the stress acts, the second the direction. In Fig. 1.2a, there are stresses on the hidden faces. These have the same subscripts as those on the opposite faces, but the stresses are in the opposite directions. Subscripts are again used to indicate special values, such as y which indicates the yield stress.

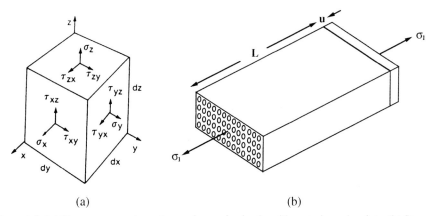

(a)   (b)

Figure 1.2. (a) Stresses on an elementary volume, dx, dy, dz, with cartesian subscripts. (b) Stress on a unidirectional fibre lamina and the corresponding extension.

In Figs. 1.1a and 1.2b, the extension u is shown. The amount of extension at a given load is proportional to the length being loaded. To indicate the effect on the material, we need to generate a value which is independent of length. This is the *strain*, which is simply the change in length divided by the original length, i.e.

$$\varepsilon = u / L \tag{1.3}$$

where ε is the symbol used for the axial tensile strain.

For shear, the strain is the angle γ, as shown in Fig. 1.1b. This is in radians, and so is dimensionless by definition. Both axial (tensile and compressive) and shear strains are subscripted. An axial strain has a single subscript for direction, corresponding to that for the stress that generates it. Shear strains have two.

There are only three shear stresses and three shear strains. This is because the order of the subscripts makes no difference. Thus $\tau_{xy} = \tau_{yx}$. (If this were not true, the vanishingly small element of material shown in Fig. 1.2a would rotate.) We thus have three possible stresses and three possible strains. Because this makes analysis difficult, the treatment is reduced to two dimensions wherever possible.

The ultimate tensile strength, σ_u, is the tensile stress required to break or fracture the material. In compression, ductile materials yield, but seldom break. Brittle materials can be broken in compression, but close examination of the fracture surfaces generally reveals that failure was initiated at some stress raiser that created a tensile stress. Thus true compression fracture is not normally seen. True shear fracture is also very rare. It may sometimes be made to occur with very ductile metals, but not with the majority of very ductile polymers (e.g. the nylons). Instead, they break in tension, sometimes at extremely high shear strains. Low molecular weight polymers with very simple structures, such as low density polythene and polypropylene can be made to fail in shear. These are the *liquid-like polymers.* Brittle polymers and ceramic materials break in tension, when sheared.

The tensile stress appears naturally in any material undergoing shear. Fig. 1b shows *simple shear.* This is not as simple as it appears; it includes rotation as well as shear. Exclude the rotation and we have *true shear.* This, as is shown in Fig 1.3, is generated by extending the material in one direction, and at the same time, squeezing it at right angles to the tension.

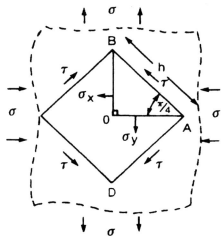

Figure 1.3. True shear generated by horizontal compression and vertical tension.

Apart from the tensile strength just described, which is also known as the *engineering tensile strength*, there are yield strengths and "true strengths". Before we define these, though, we need to discuss the stress-strain curve, which we will introduce in the next section. (Note that although there are normally no shear failure or compressive failure mechanisms which break the material into two or more pieces, there are, nevertheless, shear and compressive yield strengths.)

1.2. Yielding and the stress-strain curve

When we subject materials to constantly increasing tensile stress they will reveal a great deal about their properties, if we at the same time monitor the strain. The stress-strain plots so obtained can have many forms; two examples are shown in Fig. 1.4.

Useful metals, of which the steel and brass shown in Fig. 1.4 are good examples, give stress-strain curves which start with a linear, "elastic", region. (The region near the origin is shown, inset, with the strain expanded.) A sudden reduction of slope occurs at the *yield stress, σ_y.* This involves a slight decrease in stress with the steel shown here, but not the brass, or most metals. It can be seen that the slope decreases after the yield region, until fracture occurs, indicated by a downward arrow.

During the yielding process the bar is thinning down. For this plot we calculated the stress without taking into account this thinning. This is the *engineering stress.* When we calculate the stress based on the reduced section we obtain the *true stress.* If this is plotted instead of the engineering stress, the peak in the stress-strain curve disappears. The maximum stress is at the breaking point and gives the *true tensile strength* rather than the *engineering tensile strength.* Useful metals are generally *ductile.* This means that they can endure a high strain (typically 10% or more) before breaking.

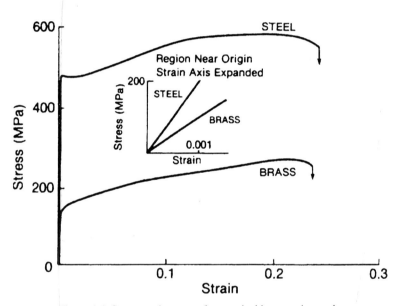

Figure 1.4. Stress-strain curves for a typical brass and a steel.

While metals have many interesting and useful properties, they are of very little relevance or interest to the composites engineer, except as ancillary materials and structures. Although much effort has gone into the making and testing of fibre reinforced metals, they are far too expensive to have a place in any normal commercial structure. Moreover, they contain self-destructive thermal stresses.

Fig. 1.5a shows the stress-strain curves for two polymers, polycarbonate and polyethylene. (A brief description of polymers will appear later – see Chapter 2.) Compare the behaviour of the polymers shown in Fig. 1.5a with the metals shown in Fig.

1.4. For polymers, the stresses are much lower and the initial slopes of the curve, in absolute terms (Pa per unit of strain) are much less. In addition, there is seldom a very sudden change of slope at the end of the initial linear region.

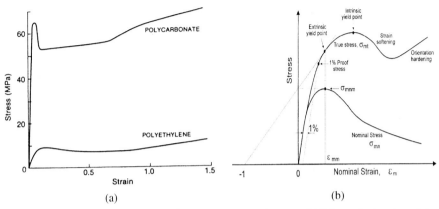

Figure 1.5. (a) Stress-strain curves for two common polymers. (b) Schematic drawing of the nominal (engineering) stress and the true stress vs. nominal strain for a polymer.

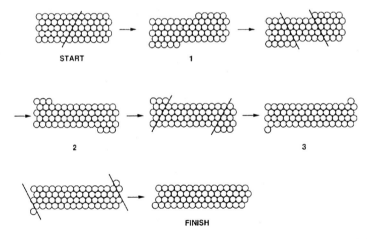

Figure 1.6. Stretching metals, at a stress greater than the yield stress, involves the sliding of the atoms over each other on the most densely packed planes. The tensile stress in this figure is acting horizontally across the page.

Metals people have a totally different definition of yield than polymer people. This is apparent from the schematic diagram shown in Fig. 1.5b. For polymers, the ex*trinsic yield point* is the peak of the nominal (engineering) stress curve and the *intrinsic yield point* is the peak of the true stress curve. (The *nominal strain* is the usual strain, as defined by Eq. (1.3). True strains are hard to evaluate.)

For a metal the 0.1% *offset stress* value is used as a measure of the yield strength. This is the stress at the point where the deviation from the initial linearity is 0.1%. It is typically at a strain of less than 0.3%. The corresponding shear stress is usually taken as the axial yield strength/$\sqrt{3}$ (von Mises criterion).

For polymers the yield strains are much higher. The *shear yield stress* of a polymer, τ_y, is defined by the equation

$$\tau_y = 1/(\; 1/ \; \sigma_{yt} + 1/ \; \sigma_{yc}) \tag{1.4}$$

where the axial intrinsic yield stresses, σ_y, have extra subscripts, t, for tension, and c, for compression. The corresponding strains are typically a few %, i.e. an order of magnitude higher than for metals. Note also that polymers have a *strain softening* region, which metals do not have. (Fig. 1.5b shows the 1% offset [or proof] stress.)

Yielding and plastic flow of polymers is different from that for metals because of their very different structures. Metal atoms can be likened to spheres which can be dragged past each other, to flow in a shear mode, as illustrated in Fig. 1.6. (This process operates through the action of extended imperfections in the structure, called dislocations.)

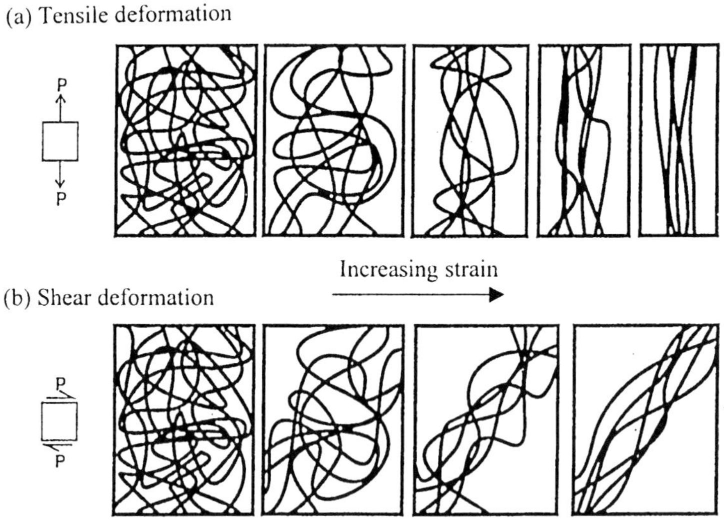

(a) Tensile deformation

Increasing strain

(b) Shear deformation

Figure 1.7. Stretching polymers in tension and shear progressively straightens the polymer chains, and orients them in the direction of the maximum tensile stress.

For polymers, the process is quite different, because the long polymer chains can be likened to pieces of string, as shown in Fig. 1.7. A tensile stress straightens and aligns the "strings", Fig. 1.7a. If a polymer is sheared, as shown in Fig. 1.7b, they straighten and align along the direction of the maximum tensile stress. (Refer to Fig. 1.3 for the relation between shear and tension.) This is why, when polymers are shear tested, they fail with an apparent shear strength which is approximately equal to the tensile strength.

1.3. Stiffnesses

The slope of the initial, straight part, of the tensile stress-strain curve is a very important property. This is the *Young's modulus, E*. It has units of Pa, but GPa are normally used. Aluminium alloys have a Young's modulus of about 71GPa, while nylons have a Young's modulus of only a few GPa, and rubbers a few MPa. There are other *stiffnesses* which are important for materials. Thus the initial slope of the stress-strain curve of a material being sheared gives the *shear modulus, G*. A material can also be compressed equally on all sides by, for example, immersing it in a fluid and compressing the fluid. In this case the change in volume per unit of volume is measured, and again the initial slope of the stress-strain curve gives a modulus, i.e. the *bulk modulus, K*.

With *isotropic materials,* which have the same properties in all directions, only two stiffnesses are required, since with two known, the third can be calculated. Solid blocks of aluminium and nylon are good examples of isotropic materials. The *Poisson's ratio, v,* provides the link. The Poisson's ratio is the ratio of the transverse shrinkage to the axial extension, measured as strains, for a bar under tension. Expresssed in terms of subscripted strains, ε_x, ε_y, for an applied tensile stress, σ_x

$$v = -\varepsilon_y / \varepsilon_x \tag{1.5}$$

Thus a simple stress analysis shows that

$$G = E / [2\,(1+v)] \tag{1.6}$$

and that

$$K = E / [3\,(1-2v)] \tag{1.7}$$

In most situations where a material is loaded, more than one stress is present. A good example of this is a beam, such as one supporting part of the floor in a building, i.e. a joist. This has significant stresses, σ_x say, along the beam's length, smaller stresses, σ_y, acting vertically, and σ_z, through the thickness. (In simple analyses of narrow beams the through thickness stresses are usually neglected.) Shears are also present, the most important one, τ_{xy}, having a strong influence on the behaviour of composite laminate beams.

The strain-stress relations are normally expressed using E and v:

$$\varepsilon_x = (\sigma_x + v\,(\sigma_y + \sigma_z))/E \tag{1.8}$$

$$\varepsilon_y = (\sigma_y + v\,(\sigma_z + \sigma_x))/E \tag{1.9}$$

$$\varepsilon_z = (\sigma_z + v\,(\sigma_x + \sigma_y))/E \tag{1.10}$$

Since there are no tensile-shear or shear-shear interactions, the shear strains, γ_{xy}, etc., are simply the shear stresses τ_{xy}, etc., divided by G.

Similar equations can be obtained relating stresses to strains by inversion. However, for anisotropic materials, things are a great deal more complicated. Here we can have 21 stiffnesses and matrices are used to relate strains to stresses, and stresses to strains. By a quirk arising from another language, the symbol C is used for the *stiffnesses* and S is used for their inverse, the *compliances*.

Fortunately, composites generally have some symmetry, so that many of these *elastic constants* are zero. Here we will describe laminates, which are normally *orthotropic* (this means they have three mutually orthogonal planes of symmetry), and have only four independent elastic constants. The compliances for a unidirectional lamina are S_{11}, S_{12}, S_{22} and S_{66}. Here the fibre axis is in the 1 direction, with the 2 direction normal to the fibres in the plane of the lamina; see Fig. 1.2b. The S_{66} refers to the shear response:

$$\begin{vmatrix} \varepsilon_1 \\ \varepsilon_2 \\ \gamma_{12} \end{vmatrix} = \begin{vmatrix} S_{11} & S_{12} & 0 \\ S_{12} & S_{22} & 0 \\ 0 & 0 & S_{66} \end{vmatrix} \cdot \begin{vmatrix} \sigma_1 \\ \sigma_2 \\ \tau_{12} \end{vmatrix} \tag{1.11}$$

It is common to use *engineering constants* instead of these compliances. When we do this we get, instead of Eqs. (1.8-10):

$$\begin{vmatrix} \varepsilon_1 \\ \varepsilon_2 \\ \gamma_{12} \end{vmatrix} = \begin{vmatrix} 1/E_1 & -v_{12}/E_1 & 0 \\ -v_{12}/E_1 & 1/E_2 & 0 \\ 0 & 0 & 1/G_{12} \end{vmatrix} \cdot \begin{vmatrix} \sigma_1 \\ \sigma_2 \\ \tau_{12} \end{vmatrix} \tag{1.12}$$

Here we have introduced subscripted Young's moduli, E_1 being for the 1 direction and E_2 for the 2 direction. G_{12} is the shear modulus in the plane of the lamina. (For isotropic materials, the reader will notice that: $E = E_1 = E_2$, $v = v_{12}$ and $G = G_{12}$.)

We need to invert this to express the stresses in terms of the strains. For this we introduce the reduced stiffnesses, Q. These have the same subscripts as the S's:

$$\begin{vmatrix} \sigma_1 \\ \sigma_2 \\ \tau_{12} \end{vmatrix} = \begin{vmatrix} Q_{11} & Q_{12} & 0 \\ Q_{12} & Q_{22} & 0 \\ 0 & 0 & Q_{66} \end{vmatrix} \cdot \begin{vmatrix} \varepsilon_1 \\ \varepsilon_2 \\ \gamma_{12} \end{vmatrix} \tag{1.13}$$

Where

$$Q_{11} = \frac{S_{22}}{S_{11}S_{22} - S_{12}^2} = \frac{E_1}{1 - v_{12}v_{21}} \tag{1.14}$$

$$Q_{12} = \frac{-S_{12}}{S_{11}S_{22} - S_{12}^2} = \frac{v_{12}E_2}{1 - v_{12}v_{21}} \tag{1.15}$$

$$Q_{22} = \frac{S_{11}}{S_{11}S_{22} - S_{12}^2} = \frac{E_2}{1 - v_{12}v_{21}} \tag{1.16}$$

and

$$Q_{66} = 1/S_{66} = G_{12} \tag{1.17}$$

These normally simplify, unless very high accuracy is required, since

$$v_{21} = v_{12}E_2/E_1 \tag{1.18}$$

For the type of composites we describe in this book E_2 can be two orders of magnitude less than E_1 and v_{12} is about 0.3, so the denominators in Eqs. (1.14-1.16) are very close to

1.00. Moreover, in practice, it is found that the constants needed for these equations cannot be estimated to better than within about 3%, due to intrinsic variability, when tests are made with five or more specimens from the same moulding. Thus, evaluating the denominator in these equations implies a spurious accuracy.

1.4. Fracture resistance

Toughness, or *fracture resistance* is a measure of the difficulty of making a crack propagate under a tensile stress (or a shear stress – but see the note at the end of this chapter). To break a material requires energy – a relatively large amount for a tough metal such as a good steel or aluminium alloy, and very little for brittle materials such as glass and other ceramics. Polymers occupy an intermediate position, but have a wide range of toughnesses, from the very low toughness of polystyrene to the relatively high toughness of polyetherether ketone (PEEK). Composite laminates are generally not very tough, and are especially easy to fracture by delamination.

The subject of fracture toughness owes its origin to A. A. Griffith, who, in 1920, noticed that specimens having identical geometry, but differing in size, had different strengths. He then went on to calculate the strain energy in a cracked material, under a tensile stress. The crack was in the shape of an ellipse, with zero width, but a finite length, a_c, which was vanishingly small in comparison to the width of the piece of material. The crack was orientated at right angles to the stress σ_∞. His celebrated equation can be written:

$$\sigma_{\infty c} = \sqrt{\frac{2E\psi}{\pi(1-v^2)a_c}} \tag{1.19}$$

where ψ is the surface energy of the material and $\sigma_{\infty c}$ the stress needed to propagate the crack. This equation is still in use, but the fracture energy term is too big to be regarded as surface energy. Nowadays we write $\mathcal{G}_{\text{I}}$ instead of 2ψ. Griffiths tested this equation with cracked glass, and found it gave the right variation with a_c, but the failure stresses predicted ($\sigma_{\infty c}$) were a bit too high. This is because there is some plastic work, even with glass. The substitution of the *work of fracture*, $\mathcal{G}_{\text{I}}$, takes care of this.

Metals and polymers are frequently tested for fracture resistance by notched impact tests, such as the Charpy Test and the Izod Test. These simple tests measure the energy required to break a specimen of the material into two pieces. More information is obtained, though, by doing *fracture toughness* tests.

Fig. 1.8a shows the single edge notched specimen used to measure the fracture toughness of a metal. Since the toughness is overestimated unless the notch is very sharp, a method has been developed to extend and sharpen the notch. To do this it is subjected to a cyclic stress. This *fatigues* the material at the notch tip, creating a fine crack with a width of little more than atomic dimensions.

Once this is done, the specimen is put under stress and its extension, or the crack opening, is monitored. When the stress vs. displacement plot starts to deviate from a straight line, crack growth has started. The stress at this point, σ_c, can be used to measure the *opening mode fracture toughness*, $\mathcal{K}_{1c}$:

$$\mathcal{K}_{1c} = \sigma_c \sqrt{\pi a}\ f(a/W) \tag{1.20}$$

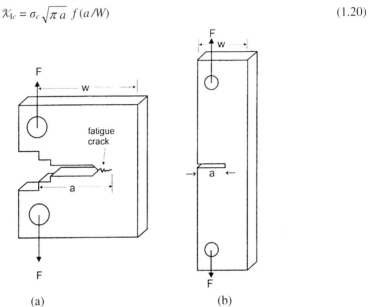

(a)   (b)

Figure 1.8. Standard specimens for measuring fracture toughness: (a) for metals and other isotropic ductile materials (ASTM E399, E1290, and E1820); (b) for quasi isotropic fibre composite laminates (ASTM E1992).

In this equation a is the length of the notch, plus its extension due to fatigue. The function $f(a/W)$ allows for the bending forces arising from the moment of F and for the finite size of the crack (relative to the specimen width W). $f(a/W)$ ranges from 8.34 for $a/W = 0.45$ to 11.36 for $a/W = 0.55$.

A similar specimen is used for laminates. It is shown in Fig. 1.8b. The notch is cut with a jeweller's saw, so is quite narrow. Extending it by fatigue is not feasible. Instead, its length is limited to between 0.5 and 0.6 of the total specimen width. $\mathcal{K}_{1c}$ is calculated from a similar formula to Eq. (1.20), but with a slight change in the function. The formula can be recast in terms of F rather than σ_f, and the maximum force is taken rather than the force at the beginning of non-linear behaviour. The formula is given with sufficient accuracy, within the allowable range of a/W values, and allowing for specimen to specimen variation, as

$$\mathcal{K}_{1c} = F_{max}(33.3\ a/W - 9.8)\ /\ [t\sqrt{W}\ ] \tag{1.21}$$

where t is the specimen thickness and W is the whole specimen width, see Fig. 1.8b, rather than the distance between the line of the force and the far specimen edge, Fig. 1.8a.

This test only works for *quasi isotropic laminates*. These are laminates which have roughly equal proportions of layers having 0, 90, +45, and -45 degree orientations relative to the main (0°) direction. This is partly because of the narrow width of the test specimen.

The fracture toughness is a measure of the more fundamental property of the material: the work of fracture, $\mathcal{G}_1$. This is the work per unit area cracked, with units of Jm^{-2}. $\mathcal{G}_1$ and $\mathcal{K}_{1c}$ are related through the elastic constants. Thus for isotropic materials

$$\mathcal{G}_1 = (1 - v^2)\,\mathcal{K}_{1c}^{\,2}/E \tag{1.22}$$

A more complicated formula is required for composite laminates because of their greater number of elastic constants. Moreover, to avoid unnecessary complication, we will drop the I subscript for the $\mathcal{K}_{1c}$, and will not use it for $\mathcal{G}$ either, since the tensile (opening) mode is the only useful one for composite laminates, and we need subscripts to define directions in these materials. (Note: the $\mathcal{G}$ used for work of fracture must not be confused with the G used for shear modulus.) For laminates, the equivalent formulae to Eq. (1.22) are given in Section 4.4.

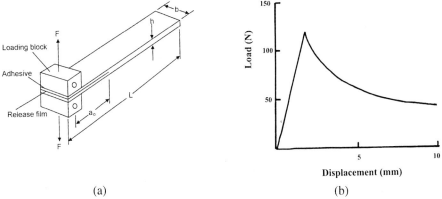

(a)   (b)

Figure 1.9. Delamination test: (a) set-up used; (b) typical force-distance plot obtained with a reinforced epoxy laminate.

Dropping a moderately heavy object, such as a spanner, on a laminate can seriously weaken it. This is because the polymers usually used to bind the fibres together are brittle; typical values for $\mathcal{G}$ for epoxies are in the range 50–200Jm^{-2}, about ten times more than for glass (about 5Jm^{-2}) but more than two orders of magnitude less than for structural aluminium alloys (10-50kJm^{-2}). Because of the brittleness, the sudden flexural forces induced in the polymer near the centre section of the sheet cause it to fracture. Since this is a much more serious problem than other modes of fracture, laminates are routinely tested for resistance to delamination, as shown in Fig. 1.9a.

The laminate has a starter crack, length a_0. This is simply a *release film*, i.e, a piece of poorly adhering polymer, such as polytetrafluoroethylene (PTFE), moulded into the laminate. The force-distance plot, Fig. 1.9b, shows a gradual build up of force, then a sudden drop, then a re-establishment of the force and a further drop, and so on. The crack grows suddenly at the force drop, and the work of delamination can be calculated from the first deviation from linearity. If this is F, and the associated displacement is δ, the work of delamination, $\mathcal{G}_3$, is determined from

$$\mathcal{G}_3 = 3F\delta/ [2ba_0] \tag{1.23}$$

where b is the specimen width and a_0 is the length of the piece of release film.

Note: We strongly recommend that **shear testing, in any form, should not be used for polymers or composites**. This is because the long chains in polymers fail by a tensile process, rather than by a shear process (see Fig 1.7). Thus **shear strength is a meaningless concept for polymers and composites**.

Further reading

S. Timoshenko: *Strength of Materials Vols. 1 & 2*, D. van Nostrand, New York, 1958.
R. N. Haward (Ed.): *The Physics of Glassy Polymers*, Applied Science Publishers, London, 1973.
S. T. Rolph and J. M. Barsom: *Fracture and Fatigue Control in Structures*, Prentice-Hall, New Jersey, 1977.

Problems

You are recommended to solve the following problems in the order given. Data needed will be found in tables in Chapter 2 and other chapters of the book.
Where a range of values is given, take the mean value.

(Note: Some of these problems, at least in the same outline form, have already appeared LBFC1 and LBFC2.)

1.1.   Calculate the breaking strains of Al alloy, high strength steel, polyvinyl chloride, glass and Si_3N_4, assuming these materials are perfectly elastic up to the breaking point.

1.2.   Calculate the weight of a rod of stainless steel, and the weight of polyethylene, both 20m long, and which when hanging vertically, will just support a man weighing 105kg, hanging on the lower end.

1.3.   A bar of material has to withstand a large temperature decrease, to a minimum temperature of 10°C, while remaining the same length. What are the corresponding maximum temperatures for bars made from Ti alloy and alumina? (Thermal expansion coefficients are: Ti alloy, 8.9MK^{-1} and alumina, 8.8MK^{-1}.)

1.4.   A ship, 165m long, is moored against a quay by a steel hawser at the bow, and a nylon rope at the stern. Both moorings are at right angles to the quayside and have a length of 9.5m and a diameter of 0.127m. They are unstressed when the ship's side is in contact with the quay. An offshore wind springs up and exerts a force of 20MN on the ship, normal to its side. What angle will the ship turn through as a result? (Note: nylon rope is stronger than bulk nylon; assume a strength of 800MPa for it.)

1.5.   A steel girder, weighing 2.6 tonnes is supported at the centre by a vertical Kevlar 149 rope 1.8cm diameter, and at the ends by vertical thermoplastic polyester ropes 1.4cm. diameter, equal in length to the Kevlar rope. Calculate the stresses in these ropes.

1.6.   A magnesium alloy rod is held at $25°$ to the vertical by a 0.28mm diameter nylon filament attached to its lower end. The upper end of the rod is held by a frictionless pivot at the same level as the upper end of the nylon filament. The rod has a diameter of 3.8cm and a length of 1.5m. If a heavy weight is hung on the rod, so that the end is pulled down, and the rod becomes approximately vertical, will the nylon be stretched to such an extent that its stress exceeds its ultimate tensile strength of 800MPa.

1.7.   A sapphire crystal disc, 3.18mm diameter and 1.05mm thick, is sandwiched between two glass sheets 0.178m in diameter. The glass is also held apart by a 1.10mm thick natural rubber ring at the periphery of the sheets, and the space between is evacuated. What would the width of the ring have to be to prevent the air pressure on the glass from breaking the sapphire, assuming the compressive strength of the sapphire is the same as the tensile strength of alumina, and that the glass doesn't break? The Young's modulus of natural rubber is 18MPa. (*The MPa is not a mistake*: the rubber is very soft.)

1.8.   Use the transformation matrix, Eq. (3.13), to show that, σ_ϕ, the stress across a plane at an angle, ϕ, to the x axis is given by

$$\sigma_\phi = \sigma_x cos^2\phi + \sigma_y sin^2\phi - 2\tau_{xy}sin\phi\,cos\phi \qquad \text{(pr 1.1)}$$

and further, that by differentiation of this equation with respect to ϕ, the maximum tensile stress in the material, $\sigma_{\phi max}$, can be determined, and this is given by

$$\sigma_{\phi max} = \frac{1}{2}\left\{\sigma_x + \sigma_y + \sqrt{(\sigma_y - \sigma_x)^2 + 4\tau_{xy}^2}\right\} \qquad \text{(pr 1.2)}$$

Use Eq. (pr 1.2) to solve the following three problems.

1.9.   A house near Bangkok was supported by four unreinforced concrete stilts. It was designed so that the concrete was under compression at a stress equal to the tensile strength of the concrete (3.5MPa). What would the speed of the wind in a typhoon have to be to cause failure of the stilts? The building presented an area of $45m^2$ to the wind, and it weighed 35 tonnes. Neglect the moment of the wind forces, and consider the stilts as subject to shear from the wind force and compression from the building weight. You may assume that the pressure exerted by the wind is ρV^2 where V is the wind velocity and ρ is the density of the air, which has a value of 1.29kg m^{-3}.

1.10. A thin walled glass tube was used to connect a motor to a stirrer in a chemical plant. The shear stress in the glass was designed to be one quarter of its tensile failure stress. Due to a design fault, the glass was also under a tensile stress along the length of the tube, and it broke. Calculate the minimum tensile load that was needed to break the tube, the diameter of which was 3.1cm and the wall thickness 1.2mm.

1.11. A link connecting a brake lever to a brake has a joint in it where two pieces of aluminium are held together by a 2.9mm diameter bolt. The bolt was brittle because of having the wrong heat treatment, and fractured at a stress of 95MPa. The bolt was screwed up tightly so that the pieces of steel were held together with a force of 300N. If the coefficient of friction of the aluminium surfaces is 0.3, what would be the maximum force the joint could transmit?

1.12. Use Eq. (pr 1.1) to show that the maximum shear stress can be determined also, and is given by

$$\tau_{\phi\theta\,max} = \frac{1}{2}\sqrt{\left(\sigma_y - \sigma_x\right)^2 + 4\tau_{xy}^2} \qquad \text{(pr 1.3)}$$

Now use Eq. (pr 1.3) to solve the following three problems, assuming that ductile materials fail on the plane of maximum shear at. $\tau_{\phi\theta max}$.

1.13. The house near Bangkok (q. 1.9) had its concrete supports replaced by steel ones. The steel was ductile and had a compressive failure stress equal to its tensile one, i.e., 150MPa. If the supports were designed to be under a compressive stress equal to one half of the material strength, what wind speed would be required to cause initiation of failure of the supports.

1.14. It was decided to use stainless steel with a tensile strength of 280MPa to replace the broken glass tube in the chemical plant (q. 1.10). The shear stress in the steel was designed to be one half the apparent shear strength of the steel. What would be the minimum load to break this tube? (Note: you must calculate the wall thickness or the cross sectional area of the tube; it has the same diameter as the glass one.)

1.15. The bolt connecting the brake lever to the brake (q. 1.11) was replaced with one that was correctly heat treated, was ductile, and had a strength of 320MPa. What would be the maximum force the joint could transmit with the new bolt?

1.16. A car is travelling along a paved country road in the spring, and has to stop suddenly at an intersection. The road has been damaged by the winter frosts, and the asphalt is broken up into pieces which are the same size as the area of contact of the car tyre on the road. It being a cold spring morning, the cracks are open near the surface, but closed further down. Determine whether the asphalt will fail, given the following data: weight of car, 1.6 tonnes; coefficient of friction of tyres on road, 1.3; apparent shear strength of asphalt, 2.2MPa; tensile strength, 0.15MPa; thickness, 2.72cm. At the instant of stopping the tyre pressure at the front is 2.2 bars, and the deceleration is sufficiently great that substantially all the car weight is supported by the front wheels. (1 bar = 100kPa.)

1.17. The materials to bridge a small crevasse, 6.2m wide, on Mt. Everest have to be as light as possible to facilitate transport. Two horizontal, parallel, wide flange beams are to be used as the main supports, since this is the most efficient shape. They must be 12m long and the cross sectional area of material in them, A, and their depth, h, must be designed so that the maximum stress does not exceed half the breaking strength. Also that under a maximum moment, M, of 30 tonne-metres on each support (due to the passage of the climbers) the radius of curvature of the beams R, has a value of 60m. The moment of area of the beam, $I = 0.7Ah^2$, the basic beam equation is $MR = EI$, and the maximum stress in the beam is $3M/[Ah]$. Develop equations for A and h, and compare the merits (so far as weight is concerned) of Al alloy, steel and Ti alloy.

1.18. A fashionable designer for a ritzy restaurant decided that a water tank could be supported by three symmetrically disposed vertical pipes, one of which could also serve for the inlet water, and the other two for the outflow. He used pewter (a tin-lead alloy) for the tank and pipes, which had a strength of 40MPa. He specified a diameter, D, for the pipes of 10.2cm and a wall thickness, t, of 1.3mm. This, he calculated, was enough with a 20% margin of safety, when the compressive stresses resulting from the weight of water in the tank were considered. He also checked to see if the pipes with this wall thickness could withstand the tensile stresses due to the water pressure. In the inlet pipe this was 0.62MPa. (The circumferential stress, σ_x, due to pressure, P, in a thin walled pipe is $\sigma_x = Pr/t$ where $2r$ is the diameter of the pipe.) This was found to be more than adequate, so he did not check the effect of the combined tensile and compressive stresses. When tested, it failed before the tank filled up. Explain why, and calculate the fraction filled at the instant of failure.

Selected answers

1.1   Al alloy, 4.37; Steel, 9.43; PVC, 1.4; Glass, 1.00; and Si_3N_4, 2.15; all millistrain.

1.3   Ti alloy, 1320°C; Alumina, 90°C.

1.5   Polyester, 1.53MPa; Kevlar, 98MPa.

1.7   0.64mm.

1.9   95ms^{-1}.

1.11   508N

1.13   74ms^{-1}.

1.15   1.14kN

1.17 Material	σ_u (MPa)	E (GPa)	ρ (Mgm^{-3})	A (cm^2)	h (mm)	W (kg)	$\sigma_u^2/E\rho$ (MNmkg^{-1})
Al alloy	310	71	2.70	93.1	62.4	302	0.50
Steel	2000	212	7.87	6.68	135	63	2.4
Ti alloy	1400	120	4.51	7.71	167	42	3.6

The Ti alloy is better than steel, which in turn scores better than the Al alloy.

Note: **You will find fracture problems at the end of Ch 2.**

Chapter 2. Material Properties

In this chapter we will give some details of properties currently achieved with metals, plastics, ceramics and fibre composites. This is to give the reader a rational basis for materials choice. It will then be appreciated that, though excellent in many ways, composites are not suitable for use in all situations.

2.1. Traditional sheets and blocks

Traditional materials, in lumps, are usually isotropic, i.e., they have the same properties irrespective of the direction chosen to make the measurements. Rolling a metal to make a sheet tends to give the crystals some preferred orientation, so that true isotropy is lost. The same is true when metals are drawn down to form wires. However, the effects are small, since the crystals of useful metals are not very anisotropic. Thus the orientation can usually be ignored.

This is not the case with polymers. These have long molecular chains, and extending and straightening them, either by rolling or drawing, can orientate the chains and drastically change their properties. Nor is this the case with ceramics. Individual crystals of ceramics can be very anisotropic. But useful forms of ceramics consist of many small crystals, usually randomly disposed. So in this section we will describe the properties of materials which do not have significant amounts of preferred orientation.

Metals comprise a large group of materials, ranging from mercury, which is molten at room temperature, tô tungsten which does not melt at 3000°C. They range from the almost totally inert gold to the highly reactive potassium. Useful metals for load bearing purposes are a small group. Moreover, they normally have to be *alloyed*, i.e. mixed with other materials, usually other metals, and often in very small quantities, to give them the good properties needed. This is because pure metals are normally soft and weak. For example, a few % of magnesium and silicon can increase the strength of aluminium from 70 to 700MPa. Less than 1% of carbon has an even bigger effect on iron.

Table 2.1 lists densities, strengths, elastic constants, works of fracture and hardnesses (measured by pressing a hard ball into the metal, under a specific load, and measuring the diameter of the resultant spherical indentation) for some typical structural metal alloys. It should be remembered that these are just values that the alloys with a particular composition and heat treatment and mechanical treatment can have. Changing the composition a little, altering the heat treatment, or doing more work hardening, can greatly change the strength, toughness and hardness.

Structural metals generally have strengths greater than 300MPa and works of fracture greater than 50kJm^{-2}. Thus magnesium and titanium alloys, which are generally not very tough, have to be used with care. The exceptionally light weight of magnesium makes it advantageous for non-critical aircraft parts, and titanium, with a low density and high temperature resistance, and also high erosion resistance, is used in aircraft (the

Concord) and for the leading edges of helicopter rotor blades. Nickel alloys are used for rotor blades in the higher temperature zones of jet engines, as also are titanium alloys.

Table 2.1. Mechanical and physical properties of some structural metals

Metal	Density (Mgm^{-3})	Strength (MPa)	E^1 (GPa)	v^2	$\mathcal{G}^3$ (kJm^{-2})	Hardness (Brinell)
Al alloy	2.7	310	71	0.345	10	590
Mg alloy	1.74	340	45	0.291	15	70
Ni alloy	8.9	930	200	0.293	400	230
High strength steel	7.87	2000	212	0.29	50	560
Stainless steel	7.9	620	215	0.29	580	170
Ti alloy	4.51	1400	120	0.361	30	200

Notes: ^1E = Young's modulus; ^2v = Poisson's ratio; $^3\mathcal{G}$ = work of fracture.
Data primarily from C. J. Smithells Metals Reference Book, fourth edition, Butterworths, London, 1967.

Metals are not good matrices for fibre composites. Countless millions of dollars have been spent on the search for a composite which might be useful at very high temperatures, to no avail. Even reinforced aluminiums have not been commercially successful. The main problem is the reactivity of the metal, but even when that can be controlled, the difference in thermal expansion between the metal and the fibre promotes the early failure of the composite when it is repeatedly heated and cooled.

Polymers also comprise a large group of materials. They contain chains or networks of atoms, chemically bonded to form molecules, which can have molecular weights of one million or more. They are usually based on carbon, with hydrogen, oxygen and nitrogen playing a major role. Compared with metals, they are weak, have low Young's moduli and are not very tough: see Table 2.2. However, they generally have excellent corrosion resistance, and are usually easy to form in final shape. They do not resist high temperatures very well: special polymers must be used if temperatures exceed 200°C, and decomposition occurs with most polymers at temperatures below 400°C.

There are two basic types of polymer: 1. thermoplastics (TP) and 2. thermosets (TS). Both types are eminently suitable for making fibre composites.

Thermoplastics have long chains of repeating chemical groups, joined together by primary chemical bonds. The backbone is often carbon, and the chains can be branched. The long chains are held together by secondary chemical bonds, and this accounts for their weakness: when the chains are straightened and aligned, the strength and modulus in the alignment direction can exceed that of a good steel.

Great diversity is achieved through different chemical composition, as well as chain length and different degrees of chain branching. Special efforts have been made to develop high temperature resistant thermoplastics for use as matrices for fibre composites. Table 2.2 includes some of these, as well as general purpose polymers and "engineering" polymers (these are generally stronger, and have lower thermal expansion).

Thermoplastics are meltable and dissolvable, in contrast to thermosets, which are neither. Thermosets are extended, three dimensional structures. Some of these can very easily be used to make fibre composites because they start out as a liquid which can be used to infiltrate the fibres. Three thermosets are listed in Table 2.2, and to ensure that it is understood that a particular chemical type of polymer can appear in both TP and TS form, this is emphasized in the table.

Table 2.2. Mechanical and physical properties of some polymers

Polymer	Density (Mgm^{-3})	Strength (MPa)	E[1] (GPa)	v[2]	$\mathscr{G}$[3] (kJm^{-2})
General purpose					
Polyethylene	0.92-93	13-28	0.26-0.52	–	–[4]
Polyvinyl chloride	1.3-1.5	41-52	2.4-4.1	–	5-10
Polystyrene	1.04-05	36-52	2.3-3.3	0.33	1.8-2.4
Engineering					
Nylon	1.12-14	76-81	1.6-3.8	0.33	2-11
Polycarbonate	1.2	66	2.4	0.37	80
Polyester[5]	1.30-38	57	1.9-3.0	–	1.3-4
Thermosets					
Epoxy	1.1-1.4	28-90	2.4-4.3	0.34	1.3-5.3
Polyester[5]	1.04-46	4 -90	2.1-4.4	0.34	1.1-2.1
Polyimide[5]	1.4	120	3.7	0.33	5-6
Special Thermoplastics					
PEEK	1.30	100	3.8	–	8.4
Polysulphone	1.24	76	2.4	–	6.3
Polyimide[5]	1.27	72-118	2.1	0.33	8

Notes: [1]E = Young's modulus; [2]v = Poisson's ratio (where a value is not shown it can be assumed that it is approximately 0.3); [3] $\mathscr{G}$ = work of fracture, [4]cannot be broken in an impact test, [5]these polymers have both thermoset and thermoplastic forms.
(Data from the Modern Plastics Encyclopedia, McGraw Hill, Highstown, NJ, published annually. ($\mathscr{G}$ values are Izod notched impact), v values are from LBFC2; special thermoplastics are from I. Y.Chang and J. K. Lees *Thermoplastic Composite Materials* 1 (1988) pp. 277-96)

(Another important type of polymer is the elastomer, or rubber. These have Young's moduli of 2–20kPa and Poisson's ratios very close to 0.50. They are not normally used for load bearing. However, they are reinforced to make tyres. But this is a major subject area in itself, and will not be included here.)

Ceramics are also important materials, but are very seldom used for carrying significant loads. This is because of their extreme brittleness. In Table 2.3 it can be seen that none even achieve a toughness of $10 Jm^{-2}$. Nevertheless, their excellent resistance to high temperatures has stimulated much research and development of high performance ceramics for use in places too hot and/or reactive for the most resistant metals. (Note: the unyielding and brittle nature of ceramics makes testing them in simple tension impractical. [They tend to get fatally crushed in the grips.] So they are normally tested in flexure. This test greatly overestimates the tensile strength.)

Table 2.3. Mechanical and physical properties of some ceramics

Ceramic	Density (Mgm^{-3})	Strength[1] (MPa)	E[2] (GPa)	v[3]	$\mathcal{G}$[4] (Jm^{-2})	Hardness $(kgmm^{-2})$
Alumina	4.0	280	400	0.20	20	2600
Glass	2.4-2.6	70	70	0.22	5	500
Glass ceramic (LAS[5])	2.4-2.6	100	100	0.24	~70	–
Magnesium oxide	3.4	200[6]	345	0.18	14[6]	330
Silicon carbide	3.2	650	455	0.19	16	2500
Silicon nitride	3.4	660[7]	320	0.22	70[7]	(very hard)

Notes: [1]flexural strength [2]E = Young's modulus; [3]v = Poisson's ratio; [4]$\mathcal{G}$ = work of fracture; [5]e.g lithium alumino silicate; [6]properties depend on grain size; [7]α-Si_3N_4 (β-Si_3N_4 has approximately half the strength and toughness).

Ceramics, like metals, are not suitable matrices for fibre composites. Again, huge efforts have gone into developing these composites without any significant successes. The only types which have been commercialised are carbon fibre reinforced carbons. However, these are not suitable for carrying significant tensile loads; they are used instead for aircraft and racing car brakes, and other situations where small loads have to be withstood at extremely high temperatures.

2.2. Filamentary forms

In about 1920, while working on the problem of the fracture of metals, A. A. Griffiths discovered that specimen size ought to have an effect on the strength of a notched specimen. So he put it to the test by making fine fibres of glass – a weak and very brittle material (see Table 2.3). They proved to be very strong; fibres with diameters of about 10 microns were more than ten times as strong as the same glass in sheet form. So was born the science of filamentary materials.

Much later it was discovered that many materials (even common salt) can be grown as very fine single crystal whiskers. These are about a tenth as thick as Griffiths' glass and can only be produced in lengths of 1mm or less, but are superlatively strong. For example a whisker of sapphire (aluminium oxide) has been reported to have a strength of 15GPa. Unfortunately, such fine whiskers are very difficult to handle, and are not normally used for reinforcements.

Glass fibres, on the other hand, are immensely successful, and this principle has been applied to other materials. These materials have been chosen because they have other desirable properties – most notably high Young's modulus and low density. Table 2.4 gives some data on some useful synthetic fibres. (Natural fibres are being used as well, but limited to applications where their lack of environmental resistance is not a problem.)

Table 2.4. Mechanical and physical properties of some reinforcing fibres

Fibre	Density (Mgm^{-3})	Diameter (μm)	Strength (MPa)	E^1 (GPa)	v^2
Kevlar[3] 49	1.44	12	3.6	130	0.35
Kevlar 149	1.47	12	2.4	160	-
Spectra[4] 900	0.97	38	2.6	120	-
Spectra 1000	0.97	27	3.0	170	-
E-glass	2.54	5-25	3.4	72	0.22
S-glass	2.48	5-15	4.8	85	-
Carbon (stiff)	1.9	5	2.3	377	0.35
Carbon (cheap)[5]	1.81	7.2	3.8	238	-
SiC (SCS6)[6]	3.32	140	4.0	420	0.19
SiC (Nicalon)	2.8	10	2.7	410	0.19
Si$_3$N$_4$	2.5	10	2.5	300	0.25
Al$_2$O$_3$ (Nextel)[7]	3.88	10-12	1.9	373	0.20

Notes: [1]Young's modulus – transverse moduli can be very much less, see fig 2.1; [2]Poisson's ratio; [3]DuPont trademark; [4]Allied Fibers trademark; [5]Zoltek; [6]AVCO; [7]3M Nextel 610.

In the table representative values are given. There is considerable variation among samples of the same fibre, and different manufacturer's products can differ significantly. The tensile properties are given. The compressive strengths of the polymer fibres can be much lower, and the moduli in compression significantly lower. *It is never safe to assume that compressive properties are the same as tensile, for any fibres.*

The polymer fibres have relatively low Young's moduli, and are not resistant to heat, but have the lowest densities. The glasses are cheap but have the lowest stiffnesses. The carbons are widely used in aerospace and sporting applications. There is great competition to provide carbon fibres, and many varieties are produced. The strongest have a strength of 5.6GPa, and the stiffest have a Young's modulus of 827GPa. Little is heard of boron nowadays, it having been replaced by silicon carbide, silicon nitride, and alumina as potential reinforcements for metals.

Apart from the glasses and some of the ceramic fibres described in Table 2.4, which are nearly isotropic, the fibres used as reinforcements have a much lower transverse Young's moduli than that in the axial direction; see Fig. 2.1. This potential weakness deserves more attention than it sometimes gets.

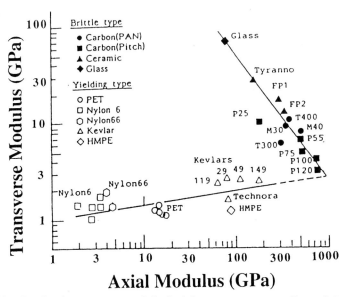

Figure 2.1. Showing that the transverse moduli of reinforcements are generally much less than the axial. (After Kawabata, S., J. Textile Institute, Vol. 81, 1990, pp. 432-7.)

2.3. Thermal properties of metals, polymers, ceramics and fibres

An important non-mechanical property is temperature resistance. Metals cover a wide range, from Mg alloys, which are only useful up to about 340°C, to Ti alloys which can be used at 1200°C, Table 2.5.

Table 2.5. Thermal expansion coefficients (CTE) and approximate maximum use temperatures of some metals and ceramics

Metal	CTE (MK^{-1})	Max. Temp. (°C)	Ceramic	CTE (MK^{-1})	Max. Temp. (°C)
Al alloy	24	350	Alumina	8.8	2045
Mg alloy	27	340	Borosilicate glass	3.5	750[1]
Ni alloy	13	1000	Glass ceramic (LAS[2])	1.5	1420
High strength steel	12	450	Magnesium oxide	3.6	2800
Stainless steel	13	750	Silicon carbide	2.6	2700
Ti alloy	10	1200	Silicon nitride	2.9	1900

Notes: [1]softening temperature   [2]lithium alumino silicate

Higher temperatures require the use of ceramics, but their brittleness at room temperature makes their employment for load bearing very risky. The differing thermal expansions among metals and ceramics (Table 2.5) leads to problems with their use in combinations, particularly with fibres potentially useful as reinforcement. This is illustrated in q. 2.17 at the end of this chapter.

The polymers are essentially low temperature materials. Also they are comparatively expansive. In Table 2.6, the glass transition temperature marks the point below which a polymer loses its softness and flexibility and becomes relatively hard and brittle.

Table 2.6. Thermal properties of some polymers

Polymer	CTE[1] (MK^{-1})	T$_g$[2] (°C)	T$_m$[3] (°C)	Polymer	CTE (MK^{-1})	T$_g$[2] (°C)	T$_m$[3] (°C)
General purpose				*Thermosets*			
Polyethylene	150±10	–122	140	Epoxy	60±5	206[4]	none
Polyvinyl chloride	62±8	90±15	–	Polyester[5]	75±25	–[4]	none
Polystyrene	67±16	90±15	–	Polyimide[5]	50±5	258[4]	none
Engineering				*Special Thermoplastics*			
Nylon	80±20	140±15	185±25	PEEK	43±3	144	340
Polycarbonate	68	155	270	Polysulphone	56	260	none
Polyester[5]	65	76±4	255±10	Polyimide[5]	50±5	258	none

Notes: [1]coefficient of thermal expansion [2]glass transition temperature; [3]melting temperature, [4]**glass transition temperatures of thermosets are normally about equal to their temperature of cure**, [5]these polymers have both thermoset and thermoplastic forms

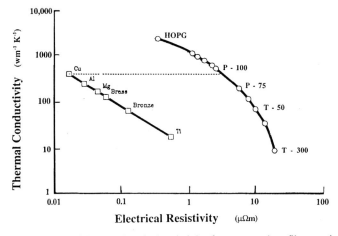

Figure 2.2. Thermal conductivity vs. electrical resistivity for some carbon fibres and some metals. (From Amoco data sheets.)

Some carbon fibres have remarkably high thermal conductivities, see Fig 2.2. Polymer fibres have negative axial CTE's, but radial CTE's comparable with those of bulk polymers; see Table 2.7.

Table 2.7. Thermal expansion coefficients (CTE) and approximate maximum use temperatures of various fibres

Fibre type	Description	Radial CTE (MK^{-1})	Axial CTE (MK^{-1})	Max. Temp. (°C)
Polyaramids:	Kevlar[1] 49	-	- 4.0	200
	Kevlar 149	59	- 2.0	200
Polyethylenes:	Spectra[2] 900	100	- 9	100
	Spectra 1000	105	- 10	100
Glasses:	E-glass	15.5	15.5	550
	S-glass	8.9	8.9	650
Carbons:	HMS (stiff)	8.0	- 0.5	2500[3]
	T300[4]	6.8	4.5	2500[3]
Polycrstalline ceramics:	SiC (SCS6)[5]	2.6	4.0	1200
	SiC (Nicalon)	3.1	~ 3.1	1300
	Si_3N_4	1.5	~ 1.5	1200
	Al_2O_3 (Nextel)[5]	7.9	~ 7.9	1200

Notes: [1]DuPont trademark; [2]Allied Fibers trademark; [3]500°C in oxidising atmosphere [4]Amoco (equivalent to Zoltec) [5]AVCO; [5]3M Nextel 610.

2.4. Unidirectional fibre laminates

Fibre composites have a wide range of properties. By choice of fibre orientation and volume fraction, even one type of fibre combined with one matrix type has the capacity to make an almost infinite variety of materials with a choice of mechanical properties limited only by the properties of the unidirectional laminate. So we will limit our description here to this form of composite. Table 2.8 lists some mechanical properties of some of them.

At present, only fibre reinforced polymers are commercially successful, so we have not listed any reinforced metals or ceramics. In any case, some reinforced polyimides can be used at temperatures up to 300°C, as much as the maximum practical use temperature of the most worthy reinforced metal: SiC-Al.

Comparing Table 2.8 with Table 2.1, we see that unidirectional reinforced polymers can match or exceed the strengths and stiffnesses of structural metals. When we take into account the lower densities, reinforced polymers remain the materials of choice for a wide range of uses, when allowance is made for poor off-axis properties. (The density of these materials can be calculated by applying the Law of Mixtures, using data from Tables 2.2 and 2.4. The volume fraction of the fibres is usually ≈ 0.7.)

Table 2.8. Mechanical properties of some unidirectional composite laminates

Fibre	Polymer	$E_1{}^1$ (GPa)	$E_2{}^2$ (GPa)	$G_{12}{}^3$ (GPa)	$v_{12}{}^4$	$\sigma_{1u}{}^5$ (GPa)	$\sigma_{2u}{}^6$ (MPa)	$\mathcal{G}^7$ (Jm^{-2})
Kevlar 49	Epoxy	76	4.0	2.1	0.34	1.4	30	≈ 10
Spectra 900	Epoxy	72	3.4	1.4	0.32	1.1	8	≈ 10
E-glass	Polyester[8]	38	10	4.0	0.26	0.7	22	low
	Epoxy	39	8.3	4.1	0.26	1.1	31	260
S-glass	Epoxy	52	-	7.6	0.28	1.6	61	130
C – stiff	Epoxy	290	5.9	4.1	0.26	0.78	47	30
C – strong	Polyimide[8]	161	9.2	-	-	2.7	66	340
	Epoxy	138	7.8	6.9	0.26	2.6	53	110
	Tough epoxy	167	8.3	-	-	2.5	64	790
	PEEK	134	9.7	4.8	0.28	2.1	92	1300

Notes: [1] axial Young's modulus; [2] transverse modulus; [3] in-plane shear modulus; [4] Poisson's ratio; [5] axial strength; [6] transverse strength; [7] work of fracture (delamination); [8] thermoset

Before we make comparisons with traditional materials, we need to make some allowance for the pronounced anisotropy of these unidirectional laminates. Even if we halve the axial properties, though, we find that these materials can still match or exceed the properties of aluminium alloys and steels, the chief metals used in structures. When we take the density into account, the advantage of reinforced polymers should be overwhelming. But they have one weakness: poor delamination resistance. Even tough polymer matrices like PEEK are not anywhere near as tough as the best aluminium alloys. The reason that reinforced polymers are fast displacing aluminium alloys is that this problem can be ameliorated by good design.

Lamination brings with it the possibility of putting together *hybrid* composites. These typically have different fibres in different layers. So polyaramid (e.g. "Kevlar") or polyethylene (e.g. "Spectra") fibres, which are comparatively tough, may be combined with carbon or glass fibres (very brittle) to make a composite which has enhanced resistance to fracture across the laminate sheet.

We will see in the next chapter how you can calculate the properties of laminates made from these raw materials.

Further reading

W. F. Smith: *Principles of Materials Science and Engineering.*, McGraw Hill, New York, 1958.

Problems

You are recommended to solve the following problems in the order given. Data needed will be found in tables in this and other chapters of the book.
Where a range of values is given, take the mean value.

(Note: Some of these problems, at least in the same outline form, have already appeared LBFC1 and LBFC2.)

2.1.   A notched fracture toughness coupon of a magnesium alloy has a width of 3.4cm and a notch length of 1.35cm. The specimen was fatigued before testing, and the notch length increased by 0.19cm. When tested, the crack started to propagate when the stress was 6.3MPa. What was the work of fracture of the material? (Hint: you can calculate an accurate enough value of f by interpolation.)

2.2.   A round bar of a high strength steel, 2.51cm diameter, has a very sharp crack in it which is 1.1mm deep. If it is loaded up till it breaks, what weight would it be supporting at the moment of fracture? (You may assume that $f = 1.0$ when the crack is very small relative to the coupon width.)

2.3.   If the bar described above was replaced by one of stainless steel with the same load carrying capacity, when both are unnotched, what weight would it support if it had the same size of notch?

2.4.   A rectangular section steel beam, used to support a small bridge, failed by brittle fracture. It was found to be incorrectly heat treated, and had a work of fracture of only $10.2 \mathrm{Jm^{-2}}$. Ultrasonic inspection revealed that failure was initiated by a crack in the lower surface of the beam. It was 0.71mm deep and 1.31cm from the centre. The beam was designed to bear a maximum stress of 220MPa. Compare the weight needed to cause failure with the design maximum weight. The weight is supported by loading the beam at its centre. The beam has a span of 4.8m, a width (b) of 5.1cm and a depth (d_1) of 21cm. (For a beam of this shape, $I = bd_1^3/12$; refer to q. 1.17 for further information about beams.)

2.5.   Show that the plastic work done, per unit volume, of a perfectly plastic yielding metal, is simply equal to the product of the stress and the strain. Hence estimate the depth of deformed material at each crack face for an aluminium alloy which has fractured with $\mathcal{G}_1 = 270 \mathrm{kJm^{-2}}$. The yield strength of the material is 120MPa, and X-ray analysis showed that the plastic strain in the deformed zone was 0.98. (Assume that, after yielding, the aluminium is perfectly plastic.)

2.6.   A rod, 0.85mm in diameter, is used to support a 5.6kg weight on a machine used to measure creep. The rod was made from brass, and was strong enough to support a weight of 16.7kg. It was, however, brittle, and failure at the maximum load was caused by flaws 6.6µm deep. Because of this brittleness, care is needed when loading the machine, so that the 5.6kg weight is not released at a level much higher than its working position. What is the maximum height that the load can fall through without the wire breaking, given that, for brass, $v = 0.350$ and $E = 101$GPa. (Note: *creep* is measured by prolonged loading, at a fraction of the breaking load, and usually at an elevated temperature. Under the correct conditions, practically any material will creep.)

2.7.   A ship's hatchway has a tiny crack in one of its corners where it is welded to the steel deck plates. If the crack extends into the deck plates for a distance of 0.81mm, and is oriented at right angles to the stress in the deck, what would the stress have to be to make it extend more, if the work of fracture of the steel is 2.65kJm^{-2}? The hatch is 4.00m square, has corners rounded to 25mm, and has one side parallel to the stress. You can use the approximate formula for stress concentration at the corner of a step or at the root of a surface crack, of length a (or a crack in the centre of length $2a$) and tip radius r_{tip}:
$\sigma_{tip} \cong 2\sigma_{\infty}\sqrt{a/r_{tip}}$. You may assume this stress falls off with distance from the hatch as the reciprocal of the square root of the distance, so that it has fallen to one half of its maximum value at a distance of 12.5mm. Is this average stress, σ_{∞} , in the deck plates, high enough for the crack to propagate right across the deck.

2.8.   Compare the maximum weights that can be supported by a Kevlar 49 fibre, a Nicalon fibre and a SCS6 fibre.

2.9.   A Nextel fibre has a surface step which is 13µm high. The radius at the inner corner of the step is 0.31µm. Will the stress at the corner reach the theoretical strength ($E/15$) at a lower applied stress than breaking strength of Nextel fibres? (See also q. 2.7.)

2.10.   The compressive strength of a Kevlar fibre is about one sixth of its tensile strength. Estimate the smallest diameter of rod on which the fibre can be wound without damage due to excessive compression. (Assume that the modulus in compression is equal to the Young's modulus.)

2.11.   The distribution of surface cracks in a production run of glass fibres is such that each 10cm length has, on average, one crack which is 1µm long, 10 cracks which are 0.1µm long, etc., so that the number of cracks of length a (µm) is $1/a$. Derive a relationship representing the strength of the fibres as a function of fibre length, and hence calculate the strength of a 2.54cm length of the fibre.

2.12.   An E-glass fibre 10.3µm diameter and 3.4cm long falls upon another fibre at a speed of 46mms^{-1}. Assume that it is stopped by the fibre that it hits, and as a result the stationary fibre is cracked. Calculate the surface area of crack produced if all the kinetic energy goes into producing the crack. If the crack has a constant depth around the whole fibre circumference, what would the strength of the fibre be reduced to as a result of the crack?

2.13   Calculate the maximum load that can be supported by a roving which consists of 1226 E-glass fibres and 15 SCS6 fibres. (Hint: check the fibre strain; and assume the glass fibres all have the same strength, and all the SCS6 fibres likewise.) The diameter of the glass fibres was 10 microns.

2.14.   What is the maximum length of a material that can be hung vertically? Compare unidirectional S-glass-epoxy, strong carbon-epoxy (both having an epoxy matrix with a density of 1.32Mgm^{-3}) with the strongest metal (Table 2.1) and the strongest polymer (Table 2.2).

2.15.   What would be the extension at the breaking point of the above lengths of material?

2.16.   A horizontal 2.56m^2 square piece of a unidirectional laminate, made from stiff carbon-epoxy with a thickness of 2.90mm, is subject to a bending moment so that it has a radius of curvature of 26cm. A wrench weighing 1.7kg is dropped on it from a height of 21cm. Assuming that all the kinetic energy is absorbed by delamination at the central plane of the laminate, what is the new radius of curvature? You may assume the laminate acts like two half thickness beams over the delaminated area, and the change in curvature is determined by the fraction delaminated. Also assume that the laminate has the same properties in flexure as it does in tension.

2.17.   It was shown in LBFC2 that the axial thermal stress, σ_{mz}, in the matrix of a unidirectional lamina of a reinforced metal, can be approximately estimated, between fibre volume fractions (V_f) of 0.3 and 0.7, from

$$\sigma_{mz} = E_m(T_2 - T_1)\,(\alpha_{fz} - \alpha_m)\,\{10V_f + V_m E_f^* + 2a\,[1 + 2V_f] - 3\}/7 \qquad \text{(pr 2.2)}$$

where T_2 and T_1 are the temperatures prior to and after cooling, α_{fz} and α_{rz} are the axial and radial CTE's of the fibres and α_m is the CTE of the matrix, $E_f^* = E_f/E_m$ and

$$a = (\alpha_{fz} - \alpha_m)/(\alpha_{fr} - \alpha_m) \qquad \text{(pr 2.3)}$$

Carbon reinforced aluminium with $V_f = 0.5$ has been made by hot pressing the matrix surrounding individual fibres at 475°C. Calculate the stress in the matrix after the composite has cooled to 20°C, assuming perfect elasticity. The pure aluminium used for the matrix has a yield stress of 70MPa; is this exceeded?

2.18.   The thermal stress for a unidirectional reinforced polymer is also linear between of 0.3 and 0.7. For a carbon-epoxy it is $\sigma_{mz} = 1.32\,E_m(T_2 - T_1)\,(\alpha_{fz} - \alpha_m)$ at $V_f = 0.3$ and $\sigma_{mz} = 1.55\,E_m(T_2 - T_1)\,(\alpha_{fz} - \alpha_m)$ at $V_f = 0.7$. The stress is developed in cooling from the T_g (above the T_g the stresses relax rapidly). Calculate the stresses in a lamina with a fibre volume fraction of 0.67, for a stiff carbon in an epoxy having a Young's modulus of 4.2GPa.

Selected answers

2.1. 2.67kJm^{-2}

2.3.   101tonnes

2.5.   1.11mm

2.7   27.4MPa; it would not extend indefinitely.

2.9   1.92GPa

2.11.   2.0GPa

2.13.   101kg

2.15.   Glass-epoxy,1.18km; C-epoxy, 0.92km; steel, 0.12km; polyimide, 0.14km

2.17.   −800MPa; but this is very much greater than the yield stress, so that the elasticity equations are no longer valid.

Chapter 3. Basic Concepts Needed for Design

We introduce here the essential information to enable you to design a fibre composite. We start from first principles, i.e. considering the unidirectional composite; how you estimate the stiffnesses and tensile strengths when the fibres go from end to end of the composite, do the same for the multidirectional laminate, and lastly introduce some design considerations.

3.1. The unidirectional composite

3.1.1. Estimating the elastic constants

We can choose the amount and type of fibre, and the fibre orientations and lengths. High performance composites are made from laminae having continuous fibres, all parallel. This is because, for the best composites, we need very long fibres, and as much fibre as possible. The polymer is merely there to hold the fibres together, without the risk of voids and potentially damaging fibre contacts. Such a composite is illustrated schematically in Fig 3.1.

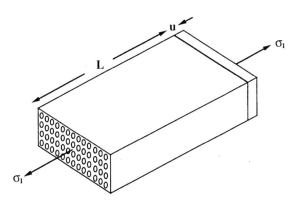

Figure 3.1. A short length of composite. The fibres are continuous and in the 1 direction.

Let this composite be stressed in the fibre direction by the stress σ_1, which extends the piece (of length L) by the amount u. The strain, ε_1, is therefore

$$\varepsilon_1 = u/L \tag{3.1}$$

We assume that the strain in the fibres and the matrix must be the same, so that the stresses in them are determined by their respective Young's moduli. Consequently the fibre stress is $\sigma_f = E_f \varepsilon_1$ and the load carried by them, assuming A is the cross section of the piece, is $AV_f E_f \varepsilon_1$, where V_f is the volume fraction of the fibres. By similar reasoning the load carried by the matrix is $AV_m E_m \varepsilon_1$, where $V_m = 1 - V_f$. (The reader will note that we have adopted the convention of subscripting the stresses, strains, moduli and volume fractions with f for the fibre, m for the matrix and 1 for the direction.)

We now add the loads in the fibres and matrix and equate this with the load on the composite, $A\sigma_1$, divide by A throughout, to get the composite stress:

$$\sigma_1 = (V_f E_f + V_m E_m)\,\varepsilon_1 \tag{3.2}$$

and defining the composite modulus in this direction by $E_1 = \sigma_1/\varepsilon_1$ gives

$$E_1 = V_f E_f + V_m E_m \tag{3.3}$$

This is the *Rule of Mixtures* for Young's modulus. It can be used with confidence: E_1 has a significant coefficient of variation (3% or more), probably due to the inevitable slight waviness of the fibres. Eq. (3.3) generally gives values comfortably within it.

When we estimate E_1, the fibres and matrix are stressed in *parallel*. But we use the *series model* for estimating the shear modulus, G_{12}. Fig. 3.2 shows this, with the fibres represented by a block, sandwiched in between sheets of matrix, and all subjected to the same shear stress τ_{12}.

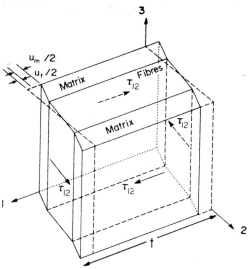

Figure 3.2. Model for estimating the shear modulus.

The fibre shear displacement, u_f, is $V_f t\, \gamma_f$ and the total matrix displacement, u_m, is equal to $V_m t\, \gamma_m$, where the γ's are the appropriate shear strains. We add these displacements, and divide by t to obtain the composite shear strain:

$$\gamma_{12} = \frac{u_f + u_m}{t} = V_f \gamma_f + V_m \gamma_m \tag{3.4}$$

Substituting $\gamma_{12} = \tau_{12}/G_{12}$, $\gamma_f = \tau_f/G_f$ and $\gamma_m = \tau_m/G_m$, therefore gives

$$\frac{1}{G_{12}} = \frac{V_f}{G_f} + \frac{V_m}{G_m} \tag{3.5}$$

This *inverse rule of mixtures* equation estimates the shear modulus with sufficient accuracy for most purposes, if it assumed the fibres and matrix behave as though they are isotropic, so that $G_f = E_f/[2(1 + v_f)]$ and $G_m = E_m/[2(1 + v_m)]$.

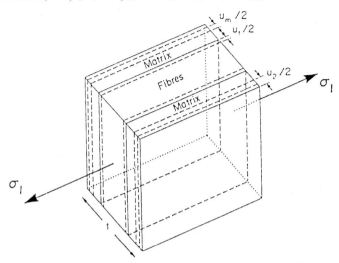

Figure 3.3. Model used for estimation of the Poisson's ratio.

To estimate the Poisson's ratio of the composite, v_{12}, we adopt the arrangement shown in Fig 3.3 This shows the model composite under a uniaxial stress, σ_1, generating a contraction in the fibres of $u_f = -v_f\varepsilon_1 V_f t$ and a corresponding contraction in the matrix of $u_m = -v_m\varepsilon_1 V_m t$. The total contraction, u_2, is the sum of these. Since the overall strain is $\varepsilon_2 = u_2/t$, we find that

$$\varepsilon_2 = -(V_f v_f + V_m v_m)\varepsilon_1 \tag{3.6}$$

and since $v_{12} = -\varepsilon_2/\varepsilon_1$, we have, finally

$$v_{12} = V_f v_f + V_m v_m \tag{3.7}$$

This serves adequately for the estimation of the composite Poisson's ratio.

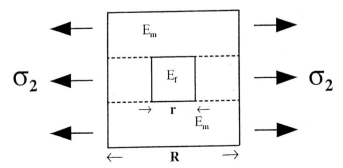

Figure 3.4. Square fibre model for transverse Young's modulus.

The transverse Young's modulus can be estimated by an analogous process, illustrated in Fig. 3.4, which shows the square fibre model. This involves matrix in series with, and parallel to the fibres.

In the centre section we have the stress exerted on the fibres and matrix in *series*, but the outer sections consist only of matrix, and are in *parallel* with the centre section. Consequently we can modify Eq. (3.3); thus,

$$E_2 = V_f^* E_f^* + (1-V_f^*)E_m \tag{3.8}$$

where V_f^* is the volume fraction occupied by the fibres in the centre section and E_f^* is its effective modulus. By inspection of Fig. 3.4 we can see that V_f^* is given by

$$V_f^* = r/R \tag{3.9}$$

For the fibres and matrix in series we assume that they are under the same stress, σ_2, as the outer regions of Fig. 3.4. We sum the fibre and matrix displacements, rather than the forces. These are given for the matrix by $(R-r)\,\sigma_2/E_m$ and the fibres by $r\sigma_2/E_{fT}$. (The transverse Young's modulus can be very different from the axial one, so we take that into account and use E_{fT} rather than E_f.) The strain is the *total displacement/R*, and 1/modulus is equal to the *total strain* $/\sigma_2$, so that we obtain for the centre section

$$\frac{1}{E_f^*} = \frac{r}{R\,E_{fT}} + \frac{R-r}{R\,E_m} \tag{3.10}$$

Now we make substitutions into Eq. (3.8). Take E_f^* from Eq. (3.10) and V_f^* from Eq. (3.9). Doing this, and eliminating r and R (since the true fibre volume fraction, V_f, is equal to r^2/R^2), gives, on simplification,

$$E_2 = E_m \left\{ 1 - \sqrt{V_f} + \sqrt{V_f} \Big/ \left(1 - \sqrt{V_f}\left[1 - E_m/E_{fT}\right]\right) \right\} \tag{3.11}$$

Overall, Eq. (3.11) comes close to estimating the correct values for glass fibre composites, which have $E_f \cong E_{fT}$, with $V_f \triangleq 0.6$, but underestimates it for $V_f \leq 0.4$. For the carbons and the polyaramids, which have $E_{fT} \approx E_f/50$, Eq. (3.11) gives results that are much too low, indicating perhaps a degree of crystallization of the polymer, increasing the effective E_m. The coefficient of variation of E_2 can be 5% or more.

If we make the assumption of a purely series arrangement (rather the series: parallel arrangement) when we estimate E_2, we obtain the inverse mixtures rule:

$$\frac{1}{E_2} = \frac{V_f}{E_f} + \frac{V_m}{E_m} \tag{3.11a}$$

This is usually accurate enough for a rough estimate of E_2, and it is not necessary to use E_{fT} instead of E_f. The reader is left to develop Eq. (3.11a) as an exercise.

Eqs. (3.3, 3.5, 3.7) and either Eq. (3.11) for glass or Eq. (3.11a) for carbon or Kevlar, provide the means for estimating the constituent properties to insert in the laminate equations that are described in Section 3.2.1.

3.1.2. Estimating the tensile strengths

We can use Fig. 3.1 to estimate the axial strength of the composite. When stretched till it breaks, we find that the composite breaking strain, ε_{1u}, is equal to the fibre breaking strain, ε_{fu}. So Eq. (3.2) becomes

$$\sigma_{1u} = (V_f E_f + V_m E_m)\,\varepsilon_{fu} \qquad\qquad (3.2a)$$

where we have written σ_{1u} for the composite axial tensile strength. Since, in high volume fraction fibre laminates such as we are dealing with here, the matrix contribution to the strength is negligible, we can use the Rule of Mixtures for this composite strength:

$$\sigma_{1u} = V_f \sigma_{fu} + V_m \sigma_{mu} \qquad\qquad\qquad\qquad (3.2b)$$

This suffices even when the matrix is so brittle that it cracks before the fibres break. In that case, we observe multiple cracks in the matrix, transverse to the fibres. But the variability in the fibre strength overwhelms the effect of the matrix contribution.

We cannot use mixture rules for the transverse strength of aligned fibre reinforced plastics. It is determined by the strength of the polymer matrix, and the strength of the bond between the fibres and the matrix. Commercially successful reinforced plastics have a very good bond between the fibres and the polymer, and measurements indicate that the transverse strength, σ_{2u}, is equal to the matrix strength, thus

$$\sigma_{2u} = \sigma_{mu} \qquad\qquad (3.12)$$

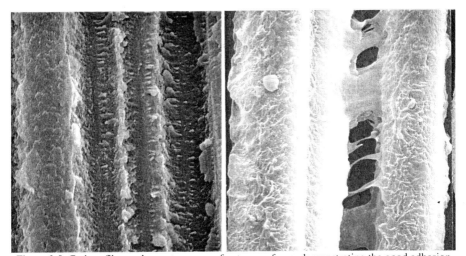

Figure 3.5. Carbon fibre-polymer transverse fracture surfaces, demonstrating the good adhesion that can be obtained with commercial moldings.

Fig. 3.5 illustrates the good adhesion that can be routinely obtained. Shown on the left is a carbon-epoxy with a transverse strength of 50MPa,. and on the right carbon-PEEK, with a transverse strength of 95MPa. They are both very nearly equal to the respective strengths of the polymer matrices.

We do not recommend trying to measure the shear strength of the fibre-matrix interface, because it appears to fail by a tensile process, and sometimes gives an inflated value for the adhesion stress. This is a corollary to the response of a polymer and a fibre composite to overwhelming shear forces described in section 4.1 hereafter.

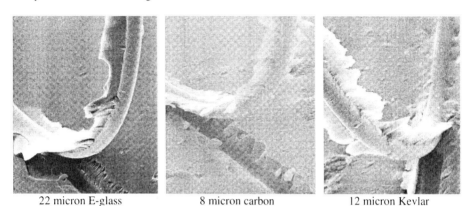

| 22 micron E-glass | 8 micron carbon | 12 micron Kevlar |

Figure 3.6. Peel testing of single fibres partially embedded in epoxy. The magnification of the pictures can be inferred from the fibre diameters.

Fig. 3.6 shows an alternative method; note how the adhesion is so good that it lifts off the polymer still adhering to the sides of the fibres.

3.1.3. Obliquely stressed composite laminae

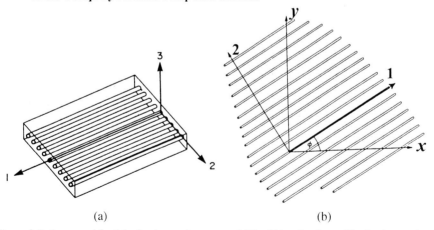

| (a) | (b) |

Figure 3.7. Axes used for (a) a lamina on its own, and (b) within a laminate. The lamina angle, ϕ, goes clockwise from the 1 direction to the x direction.

The building block of a *fibre composite laminate* is the *lamina* shown in Fig. 3.7. *Laminae* of this sort are orthotropic, so Eqs. (1.11-1.18) can be used to describe their elastic properties. Conventionally we use the 1, 2 and 3 axes as shown in Fig 3.7a.

Laminates are made by moulding laminae together, with the fibres in different directions. So we need to know what the response is, in each lamina, to oblique stresses. For this, we let the x direction be the principal direction in the composite, at angle ϕ to the fibre direction, with the y axis normal to it, as shown in Fig 3.7b. (The z direction coincides with the 3 direction, i.e. normal to the lamina.) Hence we need to carry out an axis transformation.

If the laminate experiences a general *plane stress* state, $\sigma_z \equiv \tau_{xz} \equiv \tau_{yz} \equiv 0$, as normally assumed, we can express the remaining stresses, in terms of the lamina stresses, as follows:

$$
\begin{vmatrix} \sigma_x \\ \sigma_y \\ \tau_{xy} \end{vmatrix} = \begin{vmatrix} \cos^2\phi & \sin^2\phi & -2\sin\phi\cos\phi \\ \sin^2\phi & \cos^2\phi & 2\sin\phi\cos\phi \\ \sin\phi\cos\phi & -\sin\phi\cos\phi & \cos^2\phi - \sin^2\phi \end{vmatrix} \cdot \begin{vmatrix} \sigma_1 \\ \sigma_2 \\ \tau_{12} \end{vmatrix} \tag{3.13}
$$

The same transformation applies to the strains, except the shear strains, which appear as $\gamma_{ij}/2$:

$$
\begin{vmatrix} \varepsilon_x \\ \varepsilon_y \\ \gamma_{xy}/2 \end{vmatrix} = \begin{vmatrix} \cos^2\phi & \sin^2\phi & -2\sin\phi\cos\phi \\ \sin^2\phi & \cos^2\phi & 2\sin\phi\cos\phi \\ \sin\phi\cos\phi & -\sin\phi\cos\phi & \cos^2\phi - \sin^2\phi \end{vmatrix} \cdot \begin{vmatrix} \varepsilon_1 \\ \varepsilon_2 \\ \gamma_{12}/2 \end{vmatrix} \tag{3.14}
$$

The stress-strain relationships for this set of axes is written thus:

$$
\begin{vmatrix} \varepsilon_x \\ \varepsilon_y \\ \gamma_{xy} \end{vmatrix} = \begin{vmatrix} \overline{S}_{11} & \overline{S}_{12} & \overline{S}_{16} \\ \overline{S}_{12} & \overline{S}_{22} & \overline{S}_{26} \\ \overline{S}_{16} & \overline{S}_{26} & \overline{S}_{66} \end{vmatrix} \cdot \begin{vmatrix} \sigma_x \\ \sigma_y \\ \tau_{xy} \end{vmatrix} \tag{3.15}
$$

where $|\overline{S}|$ denotes the compliance matrix for the lamina. The individual members, $\overline{S}_{ij}$, are evaluated by matrix manipulation (or can be derived directly via a lot of algebra – see LBFC2; p 109). The results are as follows.

$$\overline{S}_{11} = S_{11}\cos^4\phi + (2S_{12} + S_{66})\cos^2\phi\sin^2\phi + S_{22}\sin^4\phi \tag{3.16}$$

$$\overline{S}_{12} = S_{12}(\cos^4\phi + \sin^4\phi) + (S_{11} + S_{22} - S_{66})\cos^2\phi\sin^2\phi \tag{3.17}$$

$$\overline{S}_{22} = S_{11}\sin^4\phi + (2S_{12} + S_{66})\cos^2\phi\sin^2\phi + S_{22}\cos^4\phi \tag{3.18}$$

$$\overline{S}_{16} = (2S_{11} - 2S_{12} - S_{66})\cos^3\phi\sin\phi - (2S_{22} - 2S_{12} - S_{66})\cos\phi\sin^3\phi \tag{3.19}$$

$$\overline{S}_{26} = (2S_{11} - 2S_{12} - S_{66})\cos\phi\sin^3\phi - (2S_{22} - 2S_{12} - S_{66})\cos^3\phi\sin\phi \tag{3.20}$$

$$\overline{S}_{66} = (2[S_{11} + S_{22}] - 4S_{12} - S_{66})\cos^2\phi\sin^2\phi + S_{66}(\cos^4\phi + \sin^4\phi) \tag{3.21}$$

Alternatively we can express the $\overline{S}_{ij}$ in terms of engineering constants:

$$
\begin{vmatrix} \varepsilon_x \\ \varepsilon_y \\ \gamma_{xy} \end{vmatrix} = \begin{vmatrix} 1/E_x & -v_{xy}/E_x & \eta_{xyx}/E_x \\ -v_{xy}/E_x & 1/E_y & \eta_{xyy}/E_x \\ \eta_{xyx}/E_x & \eta_{xyy}/E_x & 1/G_{xy} \end{vmatrix} \cdot \begin{vmatrix} \sigma_x \\ \sigma_y \\ \tau_{xy} \end{vmatrix} \tag{3.22}
$$

and evaluate the terms by substituting the values of the S_{ij} from the elements of Eq (1.12). For example

$$\frac{1}{E_x} = \frac{\cos^4\phi}{E_1} + \left(\frac{1}{G_{12}} - \frac{2\nu_{12}}{E_1}\right)\cos^2\phi\sin^2\phi + \frac{\sin^4\phi}{E_2} \tag{3.23}$$

We have introduced two interaction terms, η_{xyx} and η_{xyy}, which can be evaluated in the same way.

These equations describe the elastic responses of unidirectional laminae stressed obliquely. But we need the reduced stiffnesses, $\overline{Q}_{ij}$, to describe the responses of laminates. Corresponding to Eq. (1.13) we have:

$$\begin{vmatrix} \sigma_x \\ \sigma_y \\ \tau_{xy} \end{vmatrix} = \begin{vmatrix} \overline{Q}_{11} & \overline{Q}_{12} & \overline{Q}_{16} \\ \overline{Q}_{12} & \overline{Q}_{22} & \overline{Q}_{26} \\ \overline{Q}_{16} & \overline{Q}_{26} & \overline{Q}_{66} \end{vmatrix} \cdot \begin{vmatrix} \varepsilon_x \\ \varepsilon_y \\ \gamma_{xy} \end{vmatrix} \tag{3.24}$$

where the individual terms were obtained by similar matrix manipulations as were used to evaluate the $\overline{S}_{ij}$. So we obtain analogous functions of ϕ with the Q_{ij}'s replacing the S_{ij}'s

$$\overline{Q}_{11} = Q_{11}\cos^4\phi + 2(Q_{12} + 2Q_{66})\cos^2\phi\sin^2\phi + Q_{22}\sin^4\phi \tag{3.25}$$

$$\overline{Q}_{12} = Q_{12}(\cos^4\phi + \sin^4\phi) + (Q_{11} + Q_{22} - 4Q_{66})\cos^2\phi\sin^2\phi \tag{3.26}$$

$$\overline{Q}_{22} = Q_{11}\sin^4\phi + 2(Q_{12} + 2Q_{66})\cos^2\phi\sin^2\phi + Q_{22}\cos^4\phi \tag{3.27}$$

$$\overline{Q}_{16} = (Q_{11} - Q_{12} - 2Q_{66})\cos^3\phi\sin\phi + (Q_{12} - Q_{22} + 2Q_{66})\cos\phi\sin^3\phi \tag{3.28}$$

$$\overline{Q}_{26} = (Q_{11} - Q_{12} - 2Q_{66})\cos\phi\sin^3\phi + (Q_{12} - Q_{22} + 2Q_{66})\cos^3\phi\sin\phi \tag{3.29}$$

$$\overline{Q}_{66} = (Q_{11} + Q_{22} - 2[Q_{12} + Q_{66}])\cos^2\phi\sin^2\phi + Q_{66}(\cos^4\phi + \sin^4\phi) \tag{3.30}$$

3.2. Composite laminates

We now have the tools we need to estimate the stiffnesses and discuss the strengths of composite laminates. In the interests of simplicity we will restrict our discussion to laminates which are balanced by having equal numbers of $+\phi$ and $-\phi$ layers, and have a plane of symmetry in the middle. The simplest *balanced symmetrical laminate* (conceptually at least) is made by glueing four layers together with orientations of $+\phi$, $-\phi$, $-\phi$, $+\phi$, in that order. The angle ϕ can have any value between zero and 90°. This (for $0° < \phi < 90°$) is an example of a *balanced symmetric angle ply laminate*. For brevity it is designated thus: $[\pm\phi]_s$.

Balanced symmetric laminates are the most used by far, in the high performance composites industry. This is because non-symmetric laminates have a tendency to twist when subjected to simple tension. Fig. 3.8 illustrates this for a $[\pm 25]$ laminate.

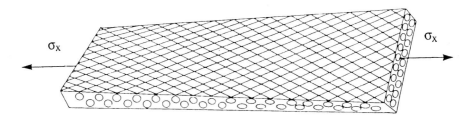

Figure 3.8. Twisting of a balanced unsymmetric 2-layer laminate.

Some forms of laminates have special names, like the *cross ply* laminate, with layers oriented at 0° and 90° to the principal direction, and the above mentioned angle ply laminate. But more clarity and precision is obtained by the almost universal adoption of an abbreviated list of orientations in square brackets. Thus

$$[90_3/\ 0_2]_{2s}$$

indicates a symmetrical cross ply laminate with three 90° layers at the surface, two 0° layers next, then three more 90° layers and two more 0° layers before we reach the central plane of the laminate. Then the sequence is reversed, so that the other surface also has three 90° layers. In the notation a comma can replace the slash.

We have already seen the application of + and – signs for angles other than 0 or 90, and the reader may have noticed the absence of the degree sign. The + sign denotes a counterclockwise rotation, as shown in Fig. 3.7.

We denote a fabric with a subscript f. For example two layers of fabric orientated at 45° to the principal direction, in the middle of a laminate, is written thus

$$[0/90/0/45_f]_s$$

and we indicate a single layer in the middle of a symmetrical laminate with a bar over the 0 or 90, whichever it is, in this way:

$$[\pm45/\ \overline{0}\ ]_s$$

3.2.1. Estimating the elastic constants

We use Classical Laminated Plate Theory, which can be expressed with apparent simplicity by matrix notation. However, this conceals a great deal of complexity. If you want to consider all its ramifications, please refer to the reading list at the end of this chapter. Here, we will only consider balanced symmetric laminates, so twisting, and the moments of forces that give rise to them, are zero.

We will consider the forces per unit length that can be applied to the edges of a laminate N_x, N_y and N_{xy}, see Fig. 3.9. Then the strains can be described in terms of the extensional stiffnesses A_{ij} thus:

$$\begin{vmatrix} N_x \\ N_y \\ N_{xy} \end{vmatrix} = \begin{vmatrix} A_{11} & A_{12} & A_{16} \\ A_{12} & A_{22} & A_{26} \\ A_{16} & A_{26} & A_{66} \end{vmatrix} \cdot \begin{vmatrix} \varepsilon_x \\ \varepsilon_y \\ \gamma_{xy} \end{vmatrix} \tag{3.31}$$

where

$$A_{ij} = \frac{1}{n} \sum_{k=1}^{n} \overline{Q}_{ij} \, (z_k - z_{k-1}) \tag{3.32}$$

and the z's are the distances from the mid-plane and n is the number of layers. For symmetric laminates, since z is positive above mid plane and negative below it, $A_{16} \equiv A_{26} \equiv 0$.

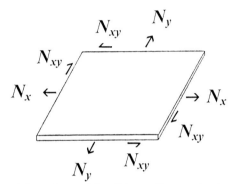

Figure 3.9. General forces (per unit length) operating on a laminate element.

The treatment is further simplified when all the laminae are made from the same thickness, t, of prepreg tape, after cure. Then $z_k - z_{k-1} = t$, and $\sigma_x = N_x/t$, $\sigma_y = N_y/t$ and $\tau_{xy} = N_{xy}/t$. So Eq. (3.31) reduces to

$$\begin{vmatrix} \sigma_x \\ \sigma_y \\ \tau_{xy} \end{vmatrix} = \begin{vmatrix} \hat{Q}_{11} & \hat{Q}_{12} & 0 \\ \hat{Q}_{12} & \hat{Q}_{22} & 0 \\ 0 & 0 & \hat{Q}_{66} \end{vmatrix} \cdot \begin{vmatrix} \varepsilon_x \\ \varepsilon_y \\ \gamma_{xy} \end{vmatrix} \tag{3.33}$$

where

$$\hat{Q}_{ij} = \frac{1}{n} \sum_{n} \overline{Q}_{ij} \tag{3.34}$$

To obtain the Engineering Constants, we need the strains in terms of stresses. We write:

$$\begin{vmatrix} \varepsilon_x \\ \varepsilon_y \\ \gamma_{xy} \end{vmatrix} = \begin{vmatrix} \hat{S}_{11} & \hat{S}_{12} & 0 \\ \hat{S}_{12} & \hat{S}_{22} & 0 \\ 0 & 0 & \hat{S}_{66} \end{vmatrix} \cdot \begin{vmatrix} \sigma_x \\ \sigma_y \\ \tau_{xy} \end{vmatrix} \tag{3.35}$$

where $\left|\hat{S}\right|$ is the compliance matrix for any symmetrical laminate. The individual members, $\hat{S}_{ij}$, are obtained by matrix inversion, as were the $\overline{S}_{ij}$ in Eq. (3.22), and may be written:

$$\hat{S}_{11} = \hat{Q}_{22}/(\hat{Q}_{11}\hat{Q}_{22}-\hat{Q}_{12}^2)=1/E_x \qquad (3.36)$$

$$\hat{S}_{12} = \hat{Q}_{12}/(\hat{Q}_{11}\hat{Q}_{22}-\hat{Q}_{12}^2) =-v_{12}/E_x \qquad (3.37)$$

$$\hat{S}_{22} = \hat{Q}_{11}/(\hat{Q}_{11}\hat{Q}_{22}-\hat{Q}_{12}^2)=1/E_y \qquad (3.38)$$

$$\hat{S}_{66} = 1/\hat{Q}_{66} = 1/G_{xy} \qquad (3.39)$$

so we can write

$$v_{xy} = \hat{Q}_{12}/\hat{Q}_{22} \qquad (3.40)$$

$$E_x = \hat{Q}_{11} - v_{xy}\hat{Q}_{12} \qquad (3.41)$$

$$G_{xy} = \hat{Q}_{66} \qquad (3.42)$$

To determine these elastic constants for a practical laminate we use Eqs. (3.25-3.27) for $\overline{Q}_{11}, \overline{Q}_{12}$ and $\overline{Q}_{22}$, together with Eq. (3.34).

We do not need E_y, because it is the complement of E_x, and can simply be obtained from

$$E_y(\phi) = E_x(90 - \phi) \qquad (3.43)$$

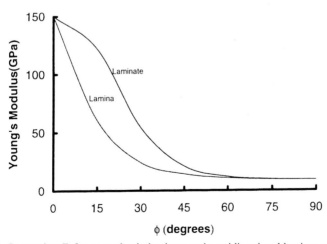

Figure 3.10. Comparing E_x for an angle ply laminate and a unidirectional lamina: carbon-epoxy with $E_1 = 150$GPa, $E_2 = 9$GPa, $G_{12} = 6$GPa and $v_{12} = 0.27$.

3.2.2. Angle ply laminates

We will now use these equations to contrast the behaviour of unidirectional and multidirectional composites. Compare a lamina, obliquely stressed at an angle ϕ, with an angle ply laminate, $[\pm\phi]_{2s}$. For these latter laminates, we simply substitute $\overline{Q}_{ij}$'s for the $\hat{Q}_{ij}$'s in Eqs. (3.40-3.42). We will plot the three important elastic constants E_x, v_{xy} and G_{xy} as ϕ goes from $0°$ to $90°$.

Fig. 3.10 shows the stiffening effect of putting laminae together in a simple $+\phi$ and $-\phi$ arrangement. E_x is greater at all angles between $0°$ and $90°$. That is true also of Poisson's ratio, Fig. 3.11, which can reach values significantly greater than 1.

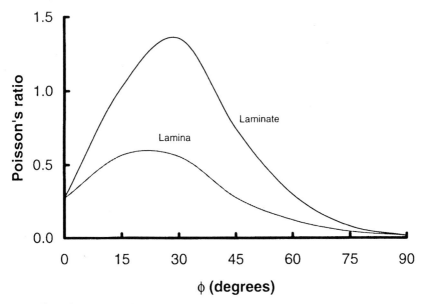

Figure 3.11. Poisson's ratios compared for the same composites as in Fig. 3.10.

In fact, Poisson's ratio can increase without apparent limit when the matrix has a sufficiently low Young's modulus compared with the fibres. Fig. 3.12 shows the effect of modulus ratio, E_1/E_2, on the maximum Poisson's ratio achievable by having a soft (low Young's modulus) matrix coupled with stiff enough fibres. (Remember E_1 and E_2 are approximately equal to V_fE_f and E_m/V_m respectively, so that $E_1/E_2 \cong V_mV_fE_f/E_m$). The laminate angle has to be judiciously chosen for this: at the top of the figure are the laminate angles.

Carbon reinforced silicone rubbers can be made with $E_1/E_2 \sim 19,000$, so a $[\pm 5]_s$ laminate made from that should have a Poisson's ratio of about 70. The line in Fig. 3.12 is given by

$$v_{xymax} = v_{12} + 0.45\sqrt{E_1/E_2} \qquad (3.44)$$

and is independent of G_{12}. The points are the calculations from lamination theory.

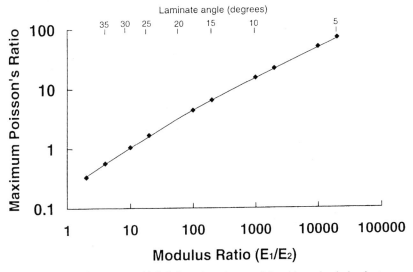

Figure 3.12. Showing the very high Poisson's ratios possible with angle ply laminates.

G_{xy} is symmetrical about 45°, Fig. 3.13, and the lamination process has even more stiffening effect than on E_x (or E_y); from about 6GPa to about 40GPa at 45°.

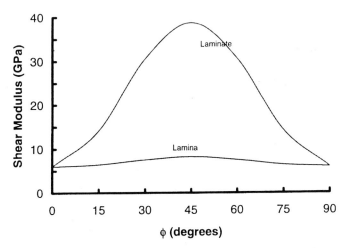

Figure 3.13. Comparing G_{xy} for an angle ply laminate and a unidirectional lamina: carbon-epoxy with $E_1 = 150$GPa, $E_2 = 9$GPa, $G_{12} = 6$GPa and $v_{12} = 0.27$ (i.e. the same as in Fig. 3.10).

3.2.3. Measuring the elastic constants

The ASTM D3039 method is usually used for the determination of E_x and the tensile strength of laminates. Whereas ductile metals and polymers are tested with a "dogbone" shaped specimen, as shown in Fig. 3.14a, composites normally have *end tabs* glued on, as shown in Fig. 3.14b. This is to protect the laminate from the damage produced by the compression exerted by the steel teeth of the testing machine grips. These end tabs sometimes have a taper to reduce stress concentrations. (Similar considerations were at work with the dogbone specimens used for metals.)

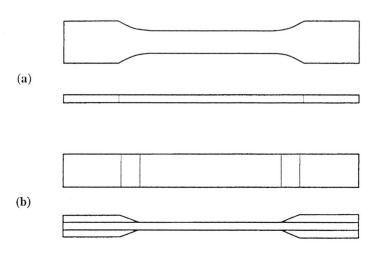

Figure 3.14. The coupon configurations used for tensile testing: a) dogbone shape for metals and polymers, b) oblong shape with end tabs for high performance composites.

The dimensions of the composite test coupon vary with the laminate lay up. For example, the recommended width for balanced and symmetric laminates is 25mm, and the length recommended is 250mm. For these specimens, the end tabs can be replaced by emery cloth. Strain gauges or extensometers are required for a valid Young's modulus measurement. ASTM does not recommend this method for laminates containing too many off axis plies because of *edge softening*.

Good alignment of the coupon is essential for accurate Young's modulus measurement. The same is true for the V-notched beam method (ASTM D5379) for the measurement of the shear modulus. Fig. 3.15 shows the test configuration. Wedges between the specimen and the gripping assembly are used to ensure good alignment. The shear modulus for $[0]_n$ and $[90]_n$ laminates, with the same composition, should the same. But more reliable results are obtained if $[0/90]_{ns}$ laminates are tested. The test is only recommended for $[0]_n$, $[90]_n$ and $[0/90]_{ns}$ laminates, or $[0/90]_n$ woven fibre or short fibre composites.

3.2.4. Strengths

The strengths of laminates are difficult to measure accurately. The ASTM methods are not suitable for measuring the strengths (or the moduli) of laminates having a significant proportion of off-axis plies, e.g. $[\pm\phi]_{2s}$, where $0° < \phi < 90°$ Pioneering attempts to make these measurements gave results which were much too low and led to the neglect of these promising forms of laminates. (The low results were due to the *disabled* fibres in the edge regions which didn't contribute to the strength.)

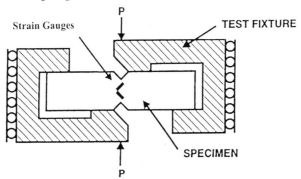

Figure 3.15. Schematic diagram of the V-notched beam method (ASTM D5379) for the measurement of the shear modulus of laminates.

Figure 3.16. Fractured $[\pm 15]_{2s}$ carbon-epoxy laminate tested in the wide coupon notched tensile test. The gauge length was 20mm and the width between the notches was 43mm.

The only reliable way of estimating the strengths of angle ply laminates are ones involving pressurized tubes. Nevertheless, having very short, wide, notched coupons comes close to giving the correct values. Fig. 3.16 shows a $[\pm 15]_{2s}$ carbon-epoxy laminate tested in this way.

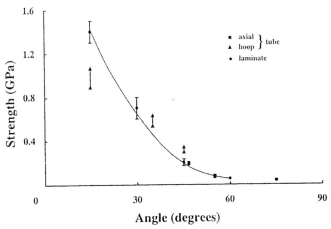

Figure 3.17. Results for glass-epoxy from wide coupon tests compared with values from tube tests.

This type of test gives results compatible with those from tube tests: see Fig. 3.17. (The tube tests are very exacting and expensive, so results are sparse. In Fig. 3.17 the vertical bars represent ±1 standard deviation.) Carbon-epoxy angle ply laminates have similar strengths. Fig. 3.18 shows the results obtained by testing coupons of these laminates with different *aspect ratios* (gauge lengths/widths). Notice in the figure how the aspect ratio affects the apparent strength. (The ASTM coupons were tested according to the 1990 version of D 3039.) The continuous line in Fig. 3.18 shows a popular theory, and indicates where the experimental results fall for very long and narrow coupons.

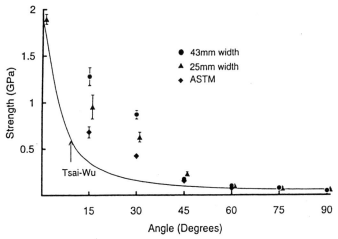

Figure 3.18. Strength results for carbon-epoxy angle ply laminates.

The modulus results obtained with these same specimens are shown in Fig. 3.19. They demonstrate the effect of edge softening and disabling of the fibres near the edges, in the narrower coupons, for the $[\pm15]_{2s}$ laminates. It is also apparent that, for the 30° and 45° angle ply laminates, the gripping arrangement is not restraining the gauge length unduly. (The gauge length is the distance between the testing grips, or in the case of end-tabbed coupons, the distance between the inner ends of the tabs.) However, the average result for the $[\pm15]_{2s}$ laminates agrees better with the upper curve than the lower. (The upper curve plots $\overline{Q}_{11}$ for this material.)

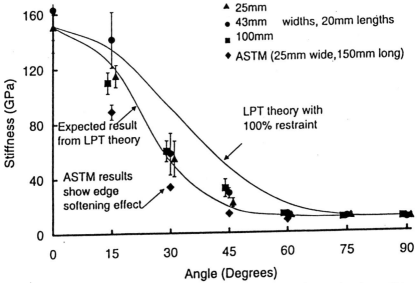

Figure 3.19. Young's modulus results for carbon-epoxy angle ply laminates of various widths and gauge lengths, compared with Laminated Plate theory.

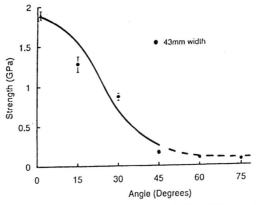

Figure 3.20. The wide coupon strengths, as shown in Fig. 3.18, fitted to $\varepsilon_{xu} = \varepsilon_{fu}$.

In this work it was observed that we had to break almost all the fibres to get something approaching the real composite strength. That could only be achieved for $\phi \leq 45°$. (The laminates containing more oblique fibres broke by a matrix dominated process, due to insufficient coupon width.) We might expect that, for $\phi \leq 45°$, the composite would break when the composite axial strain, ε_{xu}, is such that the fibres reach their breaking strain, ε_{fu}, but this leads to much too high strengths. It is better to suppose failure occurs when $\varepsilon_{xu} \cong \varepsilon_{fu}$, (i.e. $\sigma_{xu} = E_x \, \sigma_{fu} / E_f$) see Fig. 3.20. So, for the present we will take this as a working hypothesis for angle ply laminates.

Structures commonly used in the aerospace industry combine 0°, 90° and ±45° laminae in various arrangements. Although they break apart when $\varepsilon_{xu} \cong \varepsilon_{fu}$, some tranverse cracking is usually encountered as soon as $\varepsilon_{xu} = \sigma_{mu} / E_2$.

3.3. Design considerations

As touched upon in the previous paragraph, aircraft designs usually call for 0/90/±45 arrangements. These are called *pseudo isotropic* laminates, since they are very nearly isotropic in the plane of the laminate.

3.3.1. Strain limited design

The observation that, with the pseudo isotropic forms, $\varepsilon_{xu} \cong \varepsilon_{fu}$ leads naturally to the concept of strain limited design. However, the first design step is usually based on strength, allowing the full strength for the 0° layers, and 10% of the unidirectional strength for the 90° and ±45° layers. Final design is based on the maximum allowable strain, taking into consideration dynamic and expected shock loadings. The allowable strains are usually in the range 0.1–0.2%. This may seem unduly conservative, considering the strain required to break carbon fibres, but Fig. 3.21 shows the effect of having a design strain of 1.6%.

Figure 3.21. Aircraft with 1.6% strain in the wing structure. (After J. E. Gordon, Proc. Roy. Soc. (London), Vol. A282, 1964, pp. 16-23)

3.3.2. New directions in design

The almost universal use of pseudo isotropic laminates is thought to be the safest option for the current applications of high performance fibre composites. (We might refer to these more informatively, as *planar isotropic laminates*; see q. 3.13 at the end of the chapter.) But supposing we were to be more adventurous, we might chose to leave out the 90° layers which cause premature cracking and add weight. If we also leave out the 0°

layers, we have the option of an angle ply laminate for producing a stiff and strong composite. (They also can be tough – see Fig. 4.12) By choice of angle we can have a desirable stiffness also in the y direction.

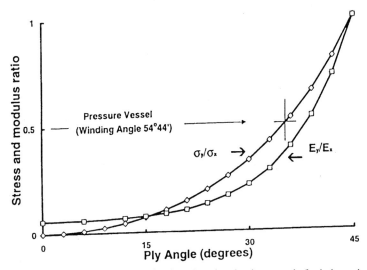

Figure 3.22. Stress ratio and modulus ratio plotted against laminate angle for balanced angle ply laminates.

Designing for stiffness with angle ply laminates also takes care of the strength. If we have our main laminate direction, x, aligned with the maximum stress, then as Fig. 3.22 shows, the ratio of stresses and stiffnesses are close to each other for $[\pm\phi]_{ns}$ angle ply laminates with ϕ between 15° and 45°. Moreover, having biaxial stresses, i.e. $\sigma_x > 0$ and $\sigma_y > 0$, is beneficial. The bursting pressure of pressure vessels, filament wound at an angle of 55.7°, where $\sigma_x = 2\sigma_y$ (see Fig 3.22), is greater than when the same tube is tested uniaxially, ($\sigma_x > 0$ and $\sigma_y = 0$).

The major design driver is the stiffness. We will make the process as general as possible by assuming biaxial stresses. Let the minimum stress be a multiple, α, of the maximum stress. We set the Principal (x) Axis in the direction of the maximum stress. (The maximum and minimum stress are normal to each other, and with axes chosen in this way, τ_{xy} is zero.)

The equations for balanced angle ply laminates are relatively simple, thus for example, we have, $E_x = \overline{Q}_{11} - v_{xy}\overline{Q}_{12}$ [Eq. (3.42) with v_{xy} given by Eq. (3.41)], and as mentioned on p. 41, E_y is the complement of E_x, and is given by

$$E_y = E_x \overline{Q}_{22} / \overline{Q}_{11} \tag{3.45}$$

The strains are

$$\varepsilon_x = \sigma_x / E_x - v_{xy}\sigma_y / E_x \tag{3.46}$$

and

$$\varepsilon_y = \sigma_y / E_y - v_{xy}\sigma_x / E_x \qquad (3.47)$$

with the shear strain being zero because we are considering the Principal Stresses.

Eliminate σ_x and σ_y between Eqs. (3.44) and (3.45) using $\sigma_y = \alpha \sigma_x$ and with

$$\varepsilon_y = \beta\varepsilon_x \qquad (3.48)$$

where β is a parameter that allows us to set the y strain independently of the x strain. This gives an equation that enables us to evaluate the ply angle, given any values of α and β:

$$\alpha E_x / E_y + v_{xy}(\alpha\beta - 1) - \beta = 0 \qquad (3.49)$$

[Suppose, for example, that we are designing an aircraft wing for a strain, which at any particular point on the laminate, is not to exceed 0.2%, for a load $P = 10$Mg per unit width in the x direction at that point. The corresponding load in the y direction, αP, is expected to be 3Mg. We can design for equal strains in the two directions, i.e. $\beta = 1$ and we have already set $\alpha = 0.3$

First we solve Eq. (3.49) by iteration using Eq. (3.45) for E_x / E_y and Eq. (3.41) for v_{xy}. This gives the ply angle, $\phi = 25.6°$, accurate within about 0.1% for the carbon-epoxy material shown in Fig. 3.18. ($E_1 = 150$GPa, $E_2 = 9$GPa, $G_{12} = 6$GPa and $v_{12} = 0.27$.) From Eq. (3.46) we can now determine the stresses, since for $\phi = 25.6°$, $E_x = 72.5$GPa, $E_y = 10.8$GPa and $v_{xy} = 1.43$. Thus $\sigma_x = 253$MPa and $\sigma_y = 76$MPa. The laminate thickness is simply the load divided by the stress, and comes to 0.39mm. We notice that the maximum stress is well below the strength, which is close to 2GPa, Fig 3.18. Since these materials are stronger under biaxial stresses, the small σ_y stress will be easily accommodated.]

Using an *organic construction* approach, we can build up our wing, area by area, based on the local loads expected. (Nature does this, as can be seen by examining the cellulose fibres in trees.) Suppose for example, another part of the wing has to withstand some other load distribution, we would probably find the principal axis (in the direction of a different maximum stress) has to be in a different direction. Thus ϕ has to take a different value, and the thickness has to be different. However, this sort of thing is easy to effect with modern methods of *fibre placement* construction (see Ch. 6).

Further reading

R. F. Gibson: *Principles of Composite Material Mechanics*, McGraw-Hill, New York, 1994

R. M. Jones: *Mechanics of Composite Materials*, McGraw-Hill, New York, 1975. (Second Edition: Taylor & Francis, Philadelphia, PA 1999.)

J. M. Whitney, I. M. Daniels and R. B. Pipes: *Experimental Mechanics of Fibre Reinforced Composite Materials*, Society for Experimental Stress Analysis, Brookfield Center, CT, 1994.

For comparison of some failure theories: Composites Science and Technology Vol. 58, 1998, pp 999-1254 and Vol. 62, 2002, pp. 1489-1514.

Problems

You are recommended to solve the following problems in the order given. Data needed will be found in tables in Chapter 2 and other chapters of the book. Where a range of values is given, take the mean value.

(Note: Some of these problems, at least in the same outline form, have already appeared LBFC1 and LBFC2.)

3.1.   Estimate the Young's modulus in the fibre direction of a unidirectional S-glass-epoxy laminate with $V_f = 0.66$.

3.2.   Estimate the stress in the fibre direction of a Kevlar 149-epoxy unidirectional laminate at a strain of 1.35%. $V_f = 0.71$.

3.3.   Estimate the axial Young's modulus of a stiff carbon-epoxy unidirectional laminate with a weight fraction of fibres, V_{fw}, of 81%., given that:
$$V_f = V_{fw} / \{ (\rho_f/\rho_m)(1 - V_{fw}) + V_{fw} \} \qquad \text{(pr 3.1)}$$
Here the ρ's are the densities of fibres and matrix.

3.4.   Calculate the maximum possible weight fraction of fibres for an alumina (Nextel)-epoxy laminate, assuming that the fibres are hexagonally packed and the laminate is void free.

3.5.   A carbon-epoxy laminate is to be designed to have a Young's modulus in the fibre direction of 56GPa. Estimate the volume fraction of the cheapest fibres that would be needed, with an epoxy having a Young's modulus of 3.3GPa.

3.6.   Calculate E_1, E_2, G_{12}, and v_{12}, for a unidirectional E-glass-epoxy laminate with $V_f = 0.72$, and hence determine the compliance matrix. The epoxy has a Young's modulus of 4.1GPa.

3.7.   Calculate the three strains ε_1, ε_2 and γ_{12} for a $V_f = 0.65$ aligned S-glass-polyester lamina stressed at right angles to the fibre direction, in the plane of the lamina. The applied stress is 51MPa and the modulus of the polyester is 3.51GPa.

3.8.   Show why the upper curve in Fig. 3.19 should be $\overline{Q}_{11}$

3.9.   A $[0/90/\pm45]_s$ carbon-epoxy laminate has constituent lamina properties of $Q_{11} = 151$GPa, $Q_{12} = 1.95$GPa, $Q_{22} = 6.47$GPa and $Q_{66} = 3.24$GPa. Estimate E_x, E_y and G_{xy} assuming that they are the mean values for 0, 90, +45 and –45 laminae, acting separately.

3.10.   Show that at least two of the above results are in error by more than 15% when lamination theory is used and the appropriate $\hat{Q}_{ij}$ are used.

3.11. Estimate E_x, E_y, v_{xy} and G_{xy} for a $[0/\pm60]_s$ laminate made from the same material as q. 3.9. How much better are these properties than the $[0/90/\pm45]_s$ above?

3.12. Calculate the angle, ϕ, for a laminate $[\pm\phi]_s$ to have a main Young's modulus, E_x, equal to that of the $[0/90/\pm45]_s$ laminate described in q. 3.9, but with properties evaluated in q. 3.10. Then go on to calculate the stress at which the first cracks might be observed, assuming that the resolved strain normal to the fibres has to be equal to the failure strain of the carbon-epoxy shown in Fig. 3.5. Also compare that with the cracking stress of the $[0/90/\pm45]_s$ laminate, calculated in the same way.

3.13. Show that the name "planar isotropic" is justified for the $[0/90/\pm45]_s$ laminate by calculating the elastic constants with the main direction, x, at angles of $22\frac{1}{2}°$ and $11\frac{1}{4}°$ to one of the lamina directions.

3.14. A rubber having a Young's modulus of 18MPa and a Poisson's ratio of 0.50 is reinforced by E-glass fibres with a volume fraction of 0.52. It is used to make an angle ply laminate with the maximum possible Poisson's ratio. Calculate the approximate Poisson's ratio that can be achieved, and estimate the angle graphically.

3.15. You are required to make a 3x motion amplifier with a glass fibre reinforced rubber laminate. Calculate the axial force, σ_x, required to force a lateral expansion, ε_y , of 10%, assuming linear behavior and laminated plate theory is upheld to this strain. The lamina from which you are required to design this has the following properties: E_1 = 43.2GPa, E_2 = 45MPa, v_{12} = 0.332, and G_{12} = 15MPa.

3.16 Estimate the strength of a 43mm wide and 20mm long coupon of a carbon-epoxy $[\pm25]_{2s}$ angle ply laminate made from the material described in q. 3.9, assuming it has been made from the cheap fibres.

3.17. Show, using Fig. 3.7b and Eq. (3.13) that, with a fibre winding angle $(90 - \phi)$ of 54° 44', the axial stress, σ_y is equal to half the hoop stress σ_x . Then go on to use the design equation, Eq. (3.49), to show that with the glass-epoxy of q. 3.6, for equal x and y strains, the winding angle needs to be somewhat greater. Calculate how much.

3.18 What difference would it make to the winding angle, if the carbon-epoxy of q. 3.9 were to be used instead of the glass-epoxy in the above question?

3.19 Design a pipe for natural gas transmission that is light, and can support its own weight over a span, l, of 30m., without breaking and with a sag not greater than 25cm. The pipe needs to withstand a pressure, P, of 1.6MPa, and not be expensive. The important stresses are the hoop stress, σ_x, generated by the gas pressure;

$$\sigma_x = PR/t,$$ (pr 3.2)

the bending stress, σ_y , given by

$$\sigma_y = 0.119A^{\frac{3}{8}}WR^{\frac{1}{4}}l^{\frac{1}{2}}t^{-1\frac{3}{4}}$$ (pr 3.3)

and the sag, z, which is

$$z = 0.0305A^{\frac{5}{8}}WR^{\frac{3}{4}}l^{1\frac{1}{2}}t^{-2\frac{1}{4}}/E$$ (pr 3.4)

Here W is the gravity force exerted by the weight, per unit of length and $A = 12 (1 - v^2)$. $2R$, the diameter of the pipe is 2.00m, t is the thickness of the pipe wall, E is the Young's modulus of the material and v the Poisson's ratio. Make your design such that the stresses never exceed half the breaking stresses. If you decide to employ a composite material, assume that the relevant Young's modulus to use for the sag is E_y. (These equations are for isotropic materials; use them anyway.)

3.20.   Eq. (1.6) can be derived from analysis of the stresses and strains applied to the element of material shown in the accompanying diagram.

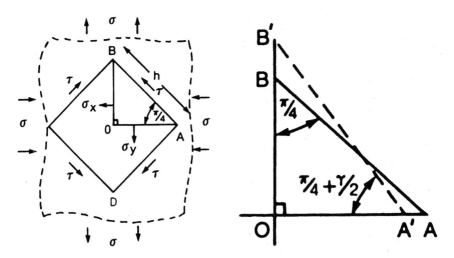

Derive the equation and then go on to apply the method to a unidirectional laminate. Compare your result with the measured shear modulus of the glass-epoxy given in Table 2.6.

3.21.   Derive an expression for the shear modulus of a symmetrical cross ply laminate by the above method with the compression and tension directed along the fibre directions. Discuss the implications for G_{12} and G_{xy}.

Selected answers

3.1.   58GPa

3.3.   279GPa

3.5.   0.225

3.7. $\varepsilon_1 = -236$ microstrain and $\varepsilon_2 = 5.5$ millistrain . Also $\gamma_{12} = 0$

3.9. $E_x = E_y = 44$GPa , $v_{xy} = 0.312$, and $G_{xy} = 4.6$GPa

3.11. $v_{xy} = 0.319$, $E_x = E_y = 55$GPa and $G_{xy} = 21$GPa

3.13. We find that $\hat{Q}_{11} = \hat{Q}_{22} = 61.16$, $\hat{Q}_{12} = 19.53$ and $\hat{Q}_{66} = 20.82$, i.e., substantially the same as in q.3.10 and 3.11.

3.15. $\phi = 30°$ and $\sigma_x = -1.20$MPa.

3.17. The winding angle, $(90 - \phi) = 63.4°$, so is about 8.7° greater.

3.19. We recommend glass-polyester filament wound pipe with winding angle of 78.7° and wall thickness of 4.75mm.

3.21. $G_{xy}^* = [Q_{11} + 2Q_{12} + Q_{22}]/4$.   Doing this we have merely approximated $G_{xy} = \overline{Q}_{66}(45°) = [Q_{11} - 2Q_{12} + Q_{22}]/4$, so we conclude that G_{12} is measured with the tension and compression directed along the diagonals of a cross ply laminate, while G_{xy} is for tension and compression along the fibres in a cross ply laminate.

Chapter 4. Laminate Resistance to Shear and Other Stressing Regimes

We have now got the tools we need to evaluate the stiffnesses, and tensile strengths but we still need to ensure that the composite will have adequate compressive strength and fracture and fatigue resistance. In this chapter we will discuss the resistance to these and other types of stress.

4.1. Shear resistance

When we subject a polymer to overwhelming shear forces, we are really testing, indirectly, the tensile strength. This is because the polymer has a "backbone" consisting of long chains of atoms, which can be regarded as some sort of elastic string. Fig. 1.7 compares the response of randomly disposed lengths of string to tension and shear. It is clear that they are the same, if you discount the rotation that accompanies the shear.

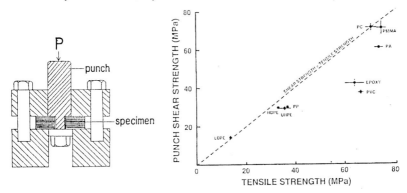

Figure 4.1. The Punch Test (ASTM D732) and the punch shear strength of some polymers compared with their tensile strengths. (After K. Liu and M.R. Piggott, Polymer Eng. Sci., Vol. 38, 1998, pp. 60-68.)

Indeed, with ductile polymers, we find that the apparent shear strength is about equal to the tensile strength. In Fig. 4.1 notice how close to the dotted line are the results for the polyethylenes (LDPE, HDPE and UHDPE), polypropylene (PP), and nylon (PA). Moreover, Fig. 4.2 shows a crack in the sheared off surface of nylon that has been punch tested. Similar cracks have been noted in UHDPE, PMMA and polycarbonate.

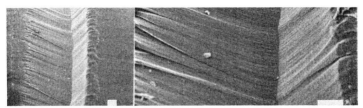

Figure 4.2. Tensile crack in nylon created by shear cutting in the punch test. White bars represent 100 microns. (After K. Liu and M.R. Piggott, Polymer Eng. Sci., Vol. 38, 1998, pp. 69-78.)

To make these observations we use the Punch Test, also shown in Fig. 4.1. Using the V Notch Test (ASTM D5379) merely stretches the polymer, as shown in Fig. 4.3. Also shown in this figure is what happens when you test more brittle polymers like polycarbonate (PC), polymethyl methacrylate (PMMA), and epoxy. These polymers break along the lines of maximum tensile stress.

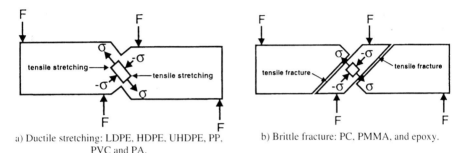

a) Ductile stretching: LDPE, HDPE, UHDPE, PP,   b) Brittle fracture: PC, PMMA, and epoxy.
PVC and PA.

Figure 4.3. V Notched Beam Test (ASTM D5379; see Fig. 3.15) applied to plastics: a) ductile polymers stretch while b) more brittle polymers break at 45°.

Not all polymers have shear strengths equal to the tensile strengths, but the overall failure process is clearly the snapping of the polymer backbone due to overwhelming tension. This is true of composites, too, and is the source of much confusion. A characteristic of shear induced failure of composites is shear hackle see Fig. 4.4. The hackle is the result of the opening up of 45° cracks formed by the strong tensile stresses. These then join together, creating leaf-like structures, as shown in Fig. 4.5.

Figure 4.4. Shear failure surface of a unidirectional laminate showing fractured fibres and shear hackle.

Figure 4.5. The origins of shear hackle lie in the tensile stresses involved in the shearing process.

The V Notch Test, suitably instrumented, is normally used to obtain a reliable estimate of the shear modulus. Shear strengths of $[0,90]_{ns}$, $[0,\pm45,90]_{ns}$ and angle ply laminates may also be measured, but as noted on page 61, there are significant orientation effects. Note also that the meaning of results from tests on unidirectional laminates is unclear.

4.2. Compressive strengths and stiffnesses

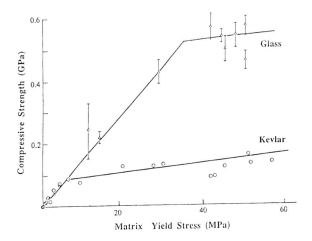

Figure 4.6. The compressive strength of a unidirectional composite is proportional to the matrix compressive yield stress, within a defined range, depending on the fibre compressive strength. (After M. R. Piggott and B. Harris, J. Materials Science, Vol. 15, 1980, pp. 2523-28.)

The compressive properties also seem to be poorly understood, despite ground-breaking work published in the early 1980's. These showed that the compressive yield stress, σ_{yc}, of the polymer matrix is an important factor controlling the compressive strength of the composite. Another factor is the waviness of the fibres. Moreover some fibres, for example the carbons, are weaker and less stiff in compression than in tension. This applies even more to the polymer fibres.

Within a certain range, depending on the fibres, the compressive strength is proportional to σ_{yc}. Fig 4.6 shows some results obtained with continuous, aligned, glass and Kevlar fibres in a polyester resin in various stages of cure. Notice that both fibres obey the same factor of proportionality near the origin. So, for the softer polymers, the axial compressive strength, σ_{1cu}, is given approximately, for this type of pultruded rod, by

$$\sigma_{1cu} \cong 5\sigma_{yc} \tag{4.1}$$

(σ_{yc} can be significantly greater than the tensile yield stress, σ_{yt}. For example for a 20°C cured epoxy σ_{yc} = 99MPa whilst σ_{yt} = 55MPa, and for PEEK the corresponding figures are 138MPa and 91MPa.)

The moduli of these polyester resins were proportional to σ_{yc}, and it was observed that the modulus of the glass-polyester pultrusion was also affected by the softness of the polymer. Fig 4.7 shows some of these results. To understand the results we need to take into account the lack of perfection of the internal structure of a practical composite, most notably the fibre waviness. (A theoretical model that explains these results is beyond the scope of this book, but is discussed in LBFC2)

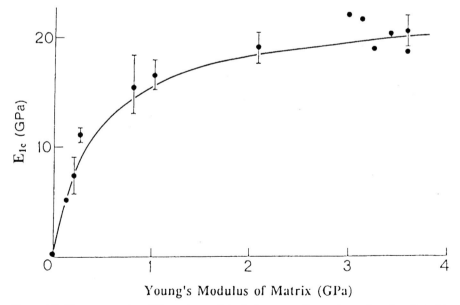

Figure 4.7. The compressive stiffness (or modulus) of a composite can be profoundly affected by the stiffness of the matrix. (After M. R. Piggott and B. Harris, as in Fig. 4.6.)

Fig. 4.6 also shows the effect of the fibre strength: the compressive strengths of polymer fibres are very much less than the tensile strengths. For Kevlar 49 it is approximately one sixth. That is the reason for the very low results for Kevlar with $\sigma_{yc} >$ 7MPa. The modulus is affected as well; with Kevlar it is approximately halved, but with increasing strain it decreases still further. The stiffness of glass is not affected by compression.

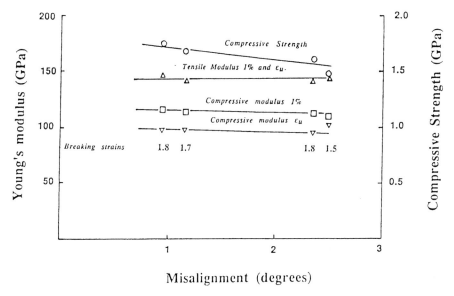

Misalignment (degrees)

Figure 4.8. Results of compact sandwich beam tests. Note that, with these unidirectional carbon fibre laminates, the Young's modulus in tension is quite a lot greater than in compression. Moreover, in compression, the modulus is strain-dependant. (After A Mrse and M. R. Piggott, Comp. Sci. Tech., Vol. 46, pp. 213-17.)

There is also evidence of a significant diminution of modulus in the case of some carbons. Fig. 4.8 shows some results with beam flexure. The compressive modulus at 1% strain is about 20% less than that in tension, and the compressive modulus at a strain of 1% is about 15% greater than at the failure strain, ε_u. Moreover, while the compressive strength is much affected by small amounts of fibre waviness, the tensile modulus is not, and the compressive modulus is barely affected. Naturally, the losses in both strength and modulus, due to fibre waviness, are much greater with laminates made from woven fibres.

Measuring the compressive properties is somewhat difficult, due to the tendency for compressed struts to buckle. For composite laminates, the specimen must be clamped in a massive fixture, such as that shown in Fig. 4.9. This fixture is used for the ASTM D3410 test method, where the load is applied through shear at the clamp surfaces. A massive fixture is also used for ASTM D6641, a method that uses both loading through the specimen ends, and shear loading. D6641 is suitable for unidirectional laminates.

Alternatively, you can use a sandwich beam, as in ASTM D5467. In this, the laminate to be tested forms the outer skins, while a honeycomb is used for the filling. This is also used for unidirectional laminates.

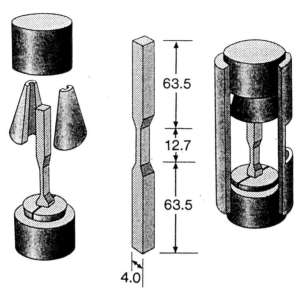

Figure 4.9.   Celanese fixture used for compression testing (ASTM D3410).

4.3. Biaxial stresses

Laminates can stand up better to biaxial tensile stresses than uniaxial tensile stresses. For example, consider ±55° E-glass-epoxy filament wound tubes. With the greater stress in the hoop direction, they are 50% stronger in 3/1 biaxial tension than in hoop tension only. A 2/1 stress ratio was also beneficial, but less so. Similar effects have been observed with carbon-epoxy also.

On the other hand, combining tension and compression weakens composite laminates. A $[0,90]_{ns}$ carbon-epoxy laminate had a strength of about 1.3GPa when tested in tension along the principal axis. Equal tension and compression directed along one of the fibre directions resulted in a failure stress of only about 0.35GPa, while when directed along the diagonals, the failure stress was further reduced to about 0.23GPa. So a state of shear is bad for laminates, and adding 45° layers, such as in $[0,\pm45,90]_{ns}$ laminates, makes them a little more resistant in some directions. Angle ply laminates don't fare any better in shear; compare Fig. 4.10 with Fig. 3.18a.

There are plenty of theoretical models purporting to predict the resistance of laminates to biaxial stresses. However there is a shortage of good data, so manufacturers are obliged to do exhaustive testing. The most reliable tests, as mentioned earlier, involve pressurised tubes (see section 3.2.4), and are difficult and expensive. In fact, there is so little confidence in the theories that, before a pressure vessel design is certified, a full scale model has to be built and tested to destruction.

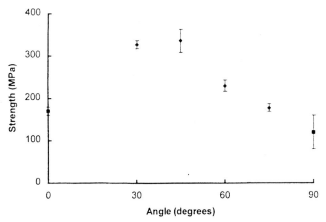

Figure 4.10. Failure under combined tension and compression (apparent shear strength) for $[\pm\phi]_{7s}$ angle ply laminates. The tension was directed 45° to the principal (x) axis. (After L. Han and M. R. Piggott, Composites, Vol. A33, 2002, pp. 35-42.)

4.4. Fracture resistance

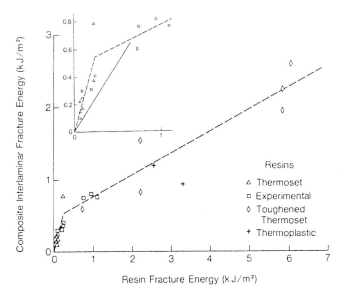

Figure 4.11. The delamination resistance of composites vs. the work of fracture of the matrix. Inset is shown a magnified view of the region near the origin. (After D. L. Hunston, Comp. Tech. Rev., Vol. 6, 1984, pp. 176-80.)

The overriding problem with laminates is that they can be split (i.e. delaminated) very easily, see Table 2.8. Typical works of fracture for delamination are 0.1–0.5kJm^{-2}. Using tough thermoplastics, e.g. PEEK, for the matrix, improves things (1.3kJm^{-2}, Table 2.8), but not anything like enough to come close to the across-the-fibres values, which are typically in the region of 1MJm^{-2}. The results of a review of works of delamination for a wide range of polymer composites is shown in Fig. 4.11.

A remedy for this problem is to have fibres normal to the plane of the laminate. Sewing laminates together has been tried with limited success (the sewing disturbs the structure and breaks some of the laminate fibres). Better, but more expensive, is to employ 3-dimensionally woven structures.

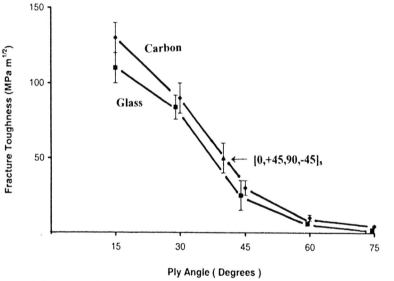

Figure 4.12. Fracture resistance of angle ply laminates, made from carbon- and glass-epoxy, compared with a quasi isotropic carbon-epoxy laminate. (After M. R. Piggott and W. Zhang, Proc. ESIS3, (Elsevier) 2002, pp. 445-54.)

The fracture' resistance of quasi isotropic $[0,\pm45,90]_{ns}$ laminates, when tested across the fibres (the *in-plane fracture toughness*), is acceptable, but angle ply laminates can be better. The fracture toughness test shown in Fig. 1.8b, while suitable for quasi isotropics, cannot be used for the angle plies. This is because the coupon isn't wide enough. So these have to be tested with a coupon similar to that shown in Fig. 3.16. Some results with this configuration are shown in Fig. 4.12.

The opening mode work of fracture, $\mathcal{G}_1$, is measured directly in delamination tests (Fig. 1.9), but the in-plane fracture toughness, $\mathcal{K}_{1c}$, is measured in the tests using the coupons shown in Fig. 1.8b and Fig. 3.16. The fracture work is the basic parameter, so we need a method to calculate $\mathcal{G}_1$ from $\mathcal{K}_{1c}$ for the in-plane fracture toughness.

For isotropic materials, we know, from Eq. (1.22), that $\mathcal{G}_1 = (1 - v^2)\,\mathcal{K}_{1c}^{\,2}/E$. We need to set out the equivalent expression for orthotropic laminates. First we have to define our symbols and realise that, with laminates, we consider plane stress instead of plane strain fracture toughness. Thus the $(1 - v^2)$ disappears, and we have

$$\mathcal{K}_j^{\,2} = E_{\mathcal{K}j}\,\mathcal{G}_j \qquad (4.2)$$

Here we use subscript j to indicate the stress direction, and subscript $\mathcal{K}$ to indicate conversion to fracture toughness. We have dropped the I indicating opening mode because that is the only mode that we will consider here. For example, a unidirectional laminate stressed in the 1 direction has $\mathcal{K}_1^{\,2} = E_{\mathcal{K}1}\mathcal{G}_1$ with

$$E_{\mathcal{K}1} = \sqrt{2} \Big/ \sqrt{S_{22}\left(\sqrt{S_{11}S_{22}} + S_{12} + S_{66}/2\right)} \qquad (4.3)$$

Similarly, a laminate stressed in the x direction has $\mathcal{K}_x$ which is related to $\mathcal{G}_x$ using Eq. (4.2), with $E_{\mathcal{K}x}$ for $E_{\mathcal{K}j}$. $E_{\mathcal{K}x}$ is calculated using Eq. (4.3), for balanced symmetrical laminates, with $\hat{S}_{ij}$ replacing the S_{ij} therein. For example, substitutions using Eqs. (3.36–3.39) gives

$$E_{\mathcal{K}x} = \frac{\sqrt{2}\left(\hat{Q}_{11}\hat{Q}_{22} - \hat{Q}_{12}^2\right)}{\sqrt{\hat{Q}_{11}\left\{\sqrt{\hat{Q}_{11}\hat{Q}_{22}} - \hat{Q}_{12} + \left(\hat{Q}_{11}\hat{Q}_{22} - \hat{Q}_{12}^2\right)/2\hat{Q}_{66}\right\}}} \qquad (4.4)$$

Fig. 4.13 shows this *fracture conversion stiffness* for angle ply laminates having different degrees of anisotropy E_1/E_2: carbon-epoxy with 16.7 and S-glass-epoxy with 5.7. (For these cases we can use $\overline{Q}_{ij}$'s instead of $\hat{Q}_{ij}$'s.) The more anisotropic carbon gives a much greater variation with ply angle.

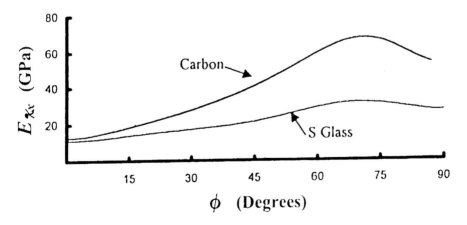

Figure 4.13. The fracture conversion stiffness, $E_{\mathcal{K}x}$, for $[\pm\phi]_{ns}$ carbon- and glass-epoxy angle ply laminates plotted vs. the angle ϕ.

The in-plane fracture toughnesses of carbon fibre laminates are in the range 10 – 100 MPa m$^{1/2}$, as shown in Table 4.1.

Table 4.1. Fracture toughness values (MPa m$^{1/2}$) for various carbon fibre laminates

Structure	Polymer	Fracture toughness
$[(0,90)_3, \overline{0}]_s$	Epoxy	19±3
$[(90,0)_3, 9\overline{0}]_s$	Epoxy	17±2
$[(0, ±60)_2, \overline{0}]_s$	Epoxy	23±4
$[90, -45, 0, +45]_{4s}$	Epoxy	57±3
$[(0,90)_3, \overline{0}]_s$	Polyimide	84±1
$[(90,0)_3, 9\overline{0}]_s$	Polyimide	45±2

(For data sources, see LBFC2, Table 10.5)

In the problems at the end of this chapter we ask you to convert some of these fracture toughnesses to works of fracture.

4.5. Fatigue resistance

Materials which are subject to alternating stresses sometimes fail at stresses which are surprisingly low compared with their failure strength when tested under conditions of monotonically increasing stress (the *quasi static strength*). The seriousness of this situation was not fully appreciated until the disastrous failures of the first commercial jet passenger plane, the British Comet. This was made of what was considered to be an adequately strong aluminium alloy, yet the fuselage broke in two in mid air. The failure started at the tip of a window, a crack propagating slowly therefrom under alternating loads, until it was so long that the material left was insufficient to support the load, so that complete failure of the whole fuselage suddenly occurred.

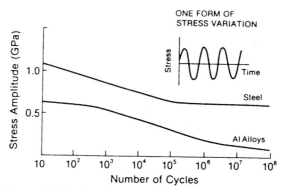

Figure 4.14. Fatigue failure curves for aluminium and steel.

Nowadays materials are routinely tested for fatigue resistance. The stressing regime can be quite complicated, if is desired to simulate service conditions. However, a good idea of the fatigue resistance can be obtained from a relatively simple test in which a piece of material is subjected to a sinusoidally varying stress (shown inset in Fig. 4.14). The stress amplitude required to break the piece is plotted as a function of the logarithm of the number of cycles it can withstand at that stress amplitude. (It is commonly referred to as the $S - N$ curve.) When this test is carried out with aluminium alloys it is found that the breaking stress decreases continuously as the number of cycles increases. Thus the material seems to have no strength after an infinite number of cycles. Steels, on the other hand, still retain about half their strength after a very large number of cycles. The difference between the two is illustrated in Fig. 4.14. The fatigue or *endurance limit* is the stress amplitude of the plateau region observed after a large number of cycles.

In metals, fatigue damage starts with plastic deformation at the surface of the specimen. Then very fine cracks begin to appear, and one of these gradually increases in length until it becomes big enough for the material to fail by the normal fracture process. The number of cycles to failure can vary greatly from specimen to specimen of the same material, even when great pains have been taken to make them identical.

Composites made with glass fibres show marked fatigue effects. This is primarily due to the long term effects of traces of moisture. When glass fibres are subject to high tensile stresses for prolonged periods, they break prematurely. This process is called *static fatigue*. Fig. 4.15 shows some endurance times for S-glass multi-fibre strands, under loads plotted as a percentage of the strength under simple loading.

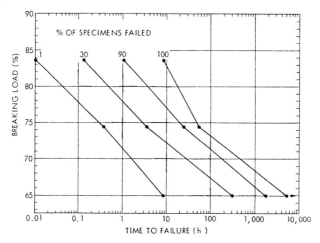

Figure 4.15. Strength remaining, expressed as a percentage of the quasi static strength (i.e. the strength under simple loading) after prolonged stressing of S-glass fibres. (After T. T. Chiao and R. L. Moore, J. Composite Materials, Vol. 5, 1971, pp. 2–11.)

Carbon fibre composites are not vulnerable to this sort of weakening effect, and have remarkably good fatigue endurance. However, the polymer may show signs of damage, especially in laminates containing 90° layers. Fig. 4.16 shows some results for carbon and S-glass.

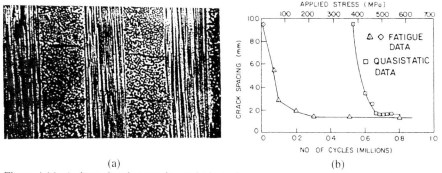

| (a) | (b) |

Figure 4.16. a) photo showing matrix cracks in an S-glass-epoxy cross ply laminate developed during fatigue, and b) crack development in 45° plies of a [0,90,±45], carbon-epoxy laminate; triangles during fatigue and squares during simple loading. (After L. J. Broutman and S. Sahu, 24th ANTEC, SPI, 1969, Section 11D, and K.L. Reifsneider. K. Schulte and J. C. Duke, ASTM STP 813, 1983, pp. 136–59.)

Fatigue endurance can be summarised in failure envelopes, such as shown in Fig. 4.17. Here there may be a constant stress superimposed on the cycling stress. Each stress may have a range of values. The continuous lines denote the values that lead to failure at a given number of cycles, e.g. 10^5 and 10^6, as shown in the figure.

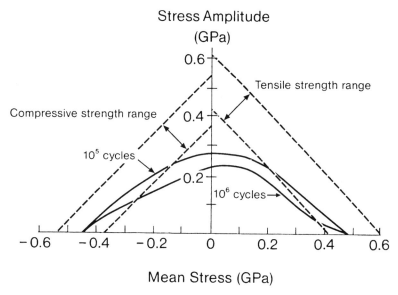

Figure 4.17. 10^5 and 10^6 cycle fatigue failure envelopes for a [0, ±45], carbon-epoxy laminate. (After L. G. Bevan, Composites, Vol. 8, 1977, pp. 227–32.)

With these laminates, compressive stresses are more damaging than tensile stresses, in line with the comparative strengths. The strains induced in the matrix are also important; thus results from one form of laminate cannot be assumed to be the same as another, because different lamination geometries are subject to different internal strains for the same applied stress.

The ratio of the [minimum stress] / [maximum stress], called the *stress ratio*, R, is an important parameter in fatigue testing. The sinusoidal stress indicated in Fig. 4.14 (inset), has $R = -1$. Note also that, as the mean stress is zero along the stress amplitude axis in Fig. 4.17, points along this line also have $R = -1$. Fig. 4.18 shows some results for unidirectional carbon-epoxy. Since fatigue failure at any stress is governed by chance, the 0.98 and 0.02 failure probabilities are shown alongside the trend line.

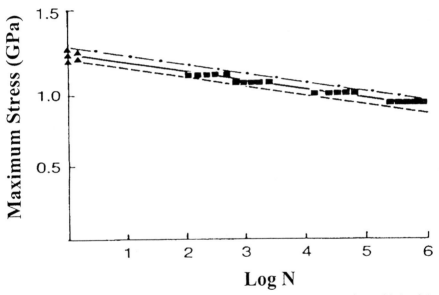

Figure 4.18. S – N curve obtained with unidirectional carbon-epoxy pultrusions with $R = 0.1$ (tension-tension fatigue). The upper and lower lines indicate 0.98 and 0.02 failure probabilities. (After P. Lam and M.R. Piggott, J. Materials Science, Vol. 24 , 1989, pp. 4427-31.)

The Young's modulus of these composites declined as they were fatigued, see Fig. 4.19. The reason is that the cyclic stress progressively damages the composite, probably making more intense the slight waviness of the fibres that is inevitably present. (Note that Fig. 4.8 shows how fibre waviness affects the modulus in compression. (These results were obtained in an exhaustive study of the origins of fatigue in unidirectional carbon-epoxy copolymers. The epoxy was copolymerized with an expanding monomer in an attempt to control the shrinkage that always accompanies the curing of the polymer.)

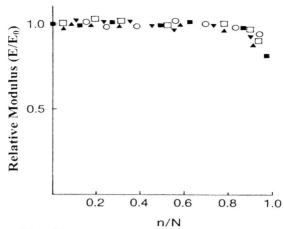

Figure 4.19. The modulus of the composite declines over the fatigue life. Here E is the Young's modulus at the nth cycle, and E_0 is the initial modulus. N is the number of cycles to failure. (After P. Lam and M.R. Piggott, J. Materials Science, Vol. 25 , 1990, pp. 1197-1202.)

Some details of the fatigue failure process have been clarified. The off-axis plies split in the early stages. Some delamination also occurs. These lead to some loss of integrity, but final failure is governed by the 0° plies. Tests with unidirectional laminates reveal that the final fatigue process is governed by the perfection of the composite. Any slight waviness of the fibres creates damaging transverse cyclic stresses, especially when the fatigue cycling involves some axial compressive stresses. This process is shown in Fig. 4.20. Composites made with woven fibres start out with pronounced waviness, and so have lower compressive strengths and are less durable in fatigue.

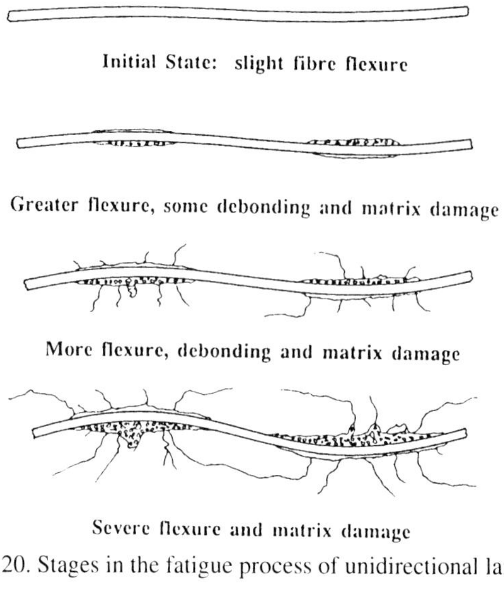

Initial State: slight fibre flexure

Greater flexure, some debonding and matrix damage

More flexure, debonding and matrix damage

Severe flexure and matrix damage

Figure 4.20. Stages in the fatigue process of unidirectional laminates.

Notches reduce the fatigue endurance. This probably reflects the increased strains in the region of the notch. Fig. 4.21 presents some results on the fatiguing of notched and drilled coupons. The more severe the notch or hole the less the endurance.

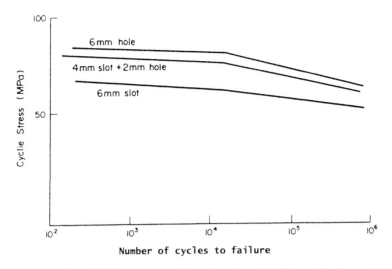

Figure 4.21. The effect of different types and size of notch on the fatigue of a glass-epoxy cross ply laminate. (After W. S. Carswell, Composites, Vol. 8, 1977, pp. 251–4.)

Further reading

In addition to the text recommended in Chapter 1, namely;

S. T. Rolph and J. M. Barsom: *Fracture and Fatigue Control in Structures*, Prentice-Hall, New Jersey, 1977

and the paper

J. G. Williams, Proc. Inst. Mech. Eng., Vol 204, 1990, pp. 209-18

Reviewing the following ASTM publications is recommended:

ASTM STP 593, 1975, ASTM STP 1110, 1991, ASTM STP 1156, 1993, ASTM STP 1230, 1995, ASTM STP 1285, 1997 and ASTM STP 1330, 1998

Problems

You are recommended to solve the following problems in the order given. Data needed will be found in tables in Chapter 2 and other chapters of the book. Where a range of values is given, take the mean value.

(Note: Some of these problems, at least in the same outline form, have already appeared LBFC1 and LBFC2.)

4.1.   Calculate the shear yield stress for the polyester showing the maximum matrix yield stress in Fig. 4.6. Assume that the intrinsic tensile yield stress is 70% of the compressive.

4.2.   Use Fig. 1.5 to estimate the intrinsic tensile yield stress for polycarbonate and polyethylene assuming no volume change and no significant differences between the intrinsic and extrinsic strains.

4.3.   Estimate the shear yield stresses of epoxy and PEEK using data given in this chapter.

4.4.   Assume that a cheap, strong carbon fibre is debonded along the whole of its length and is equivalent to a Griffiths-Irwin crack having a length equal to the fibre diameter. What would be the transverse fracture strength of the composite if the epoxy matrix had a work of fracture of $120 Jm^{-2}$, the composite had a Poisson's ratio of 0.34 and the matrix was sufficiently strong to withstand the stress? Also, what does that tell you about a single fibre debond?

4.5.   Estimate the transverse strength of a glass–epoxy composite with a fibre volume fraction of 0.71 where the tensile strength of the fibre-matrix bond is 23MPa, and the matrix strength is 67MPa. Assume hexagonal packing with close packed planes normal to the stress axis.

4.6.   The compressive strength of a pultrusion, containing 30% Kevlar 49, in an epoxy matrix with a Young's modulus of 4.1GPa, was found to be 173MPa. Estimate the compressive strength of the fibres, assuming that the epoxy is elastic up to the failure strain. (Hint: the Young's modulus of Kevlar in compression is less than that in tension.)

4.7.   If a unidirectional laminated composite could be made with perfectly straight carbon fibres, with a compressive strain to failure of 2.1% (as measured by Hahn and Sohi) what would be the failure stress in compression, given that Harper and Heumann measured the Young's modulus, E_f, as a function of strain, ε, in about 1987, with the result that can be expressed approximately by the equation $E_f = 125 + 2500\varepsilon$ GPa. The laminate has 71% of fibres in an epoxy with a Young's modulus of 3.7GPa. (Students are referred to LBFC2 for the sources of these data, and further information.)

4.8.   In a cross ply laminate the stiff carbon fibres were slightly wavy at the start of a fatigue experiment, with a wavelength, λ, of 1.6mm and an amplitude, a_0, of 10nm. It was subject to tensile-tensile fatigue at a maximum stress of 200MPa and a minimum stress of 10MPa. If the amplitude of the fibre waviness increases, under these conditions, at the rate of 0.01% per cycle, what would be the number of cycles to make the maximum fibre stress due to flexure equal to the fibre strength? The minimum radius of curvature, R_{min}, of the fibre axis is given by $R_{min} = \lambda^2/4\pi^2 a$. Comment on your result.

4.9.   A laminate coupon for a through thickness toughness test had a width of 27mm, a length of 120mm and a thickness of 4.1mm. With a notch length of 13mm, the specimen failed when the force was 2.9kN. Estimate the fracture toughness of the sample.

4.10.   The laminate in q. 4.9 was a balanced symmetrical cross ply glass-epoxy made from laminae having Q_{11} = 50GPa, Q_{22} = 10GPa, Q_{12} = 3.0GPa, and Q_{66} = 5.0GPa. Calculate the work of fracture.

4.11.   Estimate the works of fracture from the fracture toughnesses for the first three laminates listed in Table 4.1. Assume that the laminates are made from the same material as that in q. 3.9. Do the mean values directly relate to the number of zero degree layers?

4.12.   Estimate the fracture conversion stiffness for an S-glass-epoxy $[\pm30]_s$ laminate assuming Rules of Mixtures for E_1 and G_{12}, etc., with V_f = 0.65 and E_m = 2.5GPa. Hence estimate $\mathcal{G}_{fx}$ for $\mathcal{K}_x$ = 57MPa$\sqrt{m}$.

4.13.   Do the same calculations as per q. 4.12 for carbon-epoxy $[0/+45/90/-45]_s$ and hence calculate $\mathcal{G}_{fx}$ for $\mathcal{K}_x$ = 18MPa$\sqrt{m}$. Assume the laminae properties given in Fig. 3.10.

4.14.   In a delamination test on carbon-epoxy a 25mm wide beam was used which was 4.2mm thick with a release film in it which was 51mm long. When the applied force reached 175N, the displacement deviated from linearity and was 1.4mm at this point. Estimate the work of fracture.

4.15.   The same test as in q. 4.14 was carried out with carbon fibres in an unknown polymer matrix. This time the force was 385N and the displacement was 4.4mm when it deviated from linearity. Use these data and Fig 4.11 to determine what type of polymer it was.

4.16.   Estimate the minimum and mean stresses at the four levels of maximum stress indicated in Fig. 4.18.

4.17.   Identify the lines on failure envelopes, such as Fig. 4.17, which define stress ratios (R) of $-\infty$, -1, 0 and 1.

4.18.   The fatigue endurance of a material can be gauged by the average slope, β (percent per decade), of the S–N curve. Compare aluminium and steel (Fig. 4.14), carbon-epoxy (Fig. (4.18) and slotted (6mm) glass-epoxy (Fig. 4.21) on that basis.

Selected answers

4.1.   23.5MPa

4.3. τ_y (epoxy) = 35MPa
        τ_y (PEEK) = 55MPa

4.5. σ_{2u} = 40MPa

4.7. σ_{1cu} = 1.09GPa

4.9.   27MPa$\sqrt{m}$

4.11.

Laminate	E_{xx} (GPa)	$\mathscr{G}_{Ix}$ (kJm^{-2})	# of [0]'s
$[(0,90)_3, \overline{0}]_s$	29.6	12±4	7
$[(90,0)_3, \overline{90}]_s$	31.9	9±2	6
$[(0, \pm60)_2, \overline{0}]_s$	51.2	10±4	5

Looking at the table we can see that there is no simple correlation between $\mathscr{G}_{Ix}$ and the number of 0° layers.

4.13.   6.1kJm^{-2}

4.15.   2.24kJm^{-2} ;   this is a thermoplastic, or a toughened thermoset

4.17. $R = -\infty$: the line through the origin at 45° to the mean stress (S_{mean}) axis
         $R = -1$: this is the stress amplitude axis
         $R = 0$: the line through the origin at 135° to the S_{mean} axis.
         $R = 1$: this is the S_{mean} axis.

Chapter 5. Environmental Effects

Composites normally have to perform in less than ideal conditions. In this chapter we will discuss the resistance of a composite to the effects of heat, humidity and other aggressive environmental agents.

5.1. Radiation, heat and residual stress

Ultraviolet radiation can degrade the polymer matrix. External aircraft parts and structures are routinely coated with a thin aluminium films, so that such radiation is reflected rather than absorbed.

Polymers are easily degraded by heat: information about the thermal expansion, the glass transition temperature, T_g, and the melting temperature, T_m, are given in Table 2.6. Some fibres, most notably the polyaramids and polyethylenes, are also affected at relatively low temperatures, as indicated in Table 2.7. Special thermoplastics and thermosets have been developed to push the useful temperature up to about 300°C, but that appears to be about the limit. Moreover, the higher temperature polymers are very difficult to mould without air inclusions, which can weaken the composite.

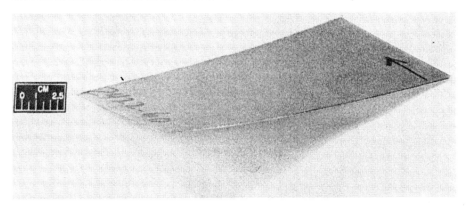

Figure 5.1. The warpage due to differential thermal shrinkage is best demonstrated with an unsymmetric laminate, as shown in this picture. Such laminates are used to measure the internal stresses. (After C.C. Chamis, Proc. ICCM2, 1978, pp. 221-39. Courtesy of the Metallurgical Soc., AIME.)

Unfortunately polymers do not necessarily fare well at low temperatures either. Polymers can be very brittle below T_g. Many epoxies used as matrix cannot be strained to the fibre breaking strain without cracking, even at room temperature. So this sets a limit to their usefulness. This is particularly serious with comparatively low modulus fibres like E-glass and S-glass. So special thermoplastics and thermosets have been developed with high strains to failure. PEEK is especially notable for being tough and not prone to this *transverse cracking* effect.

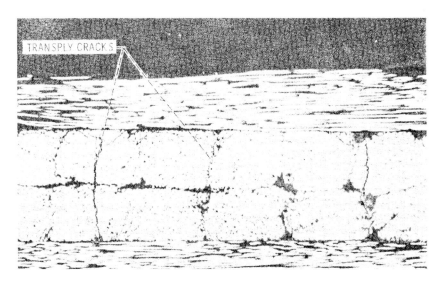

Figure 5.2. Transverse cracking produced in manufacture of a carbon-epoxy laminate. (After C.C. Chamis, Proc ICCM2, 1978 pp. 221-39. Courtesy of the Metallurgical Soc., AIME.)

The thermal expansibility of the polymers can be a problem also. It is the cause of *residual stresses* (locked-in stresses) in composites, leading to undesirable effects like warping, Fig 5.1 and cracking, Fig 5.2. The analysis of the thermal stresses is rather complicated, so we will now consider this in detail.

5.1.1 Thermal microstresses

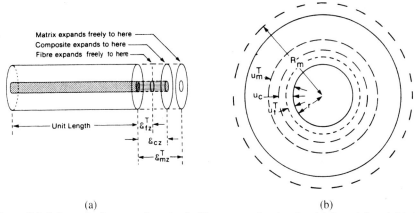

(a)   (b)

Figure 5.3. Schematic diagram of a unit of a fibre composite showing (a) the axial, and (b) the radial displacements

Thermal stresses can be treated approximately by considering the composite as being made up a regular array of repeating units, each consisting of one fibre, diameter

$2r$, surrounded by a tube of matrix, radius R_m, as shown in Fig. 5.3. For maximum simplicity in this necessarily complex subject, we make allowance for the thermal anisotropy but neglect the elastic anisotropy of the fibres.

We use cylindrical polar co-ordinates, and for the fibres, replace the x subscript by fr, the y by $f\theta$ and the z by fz. Similarly, for the matrix we use mr, $m\theta$, and mz.

We consider continuous fibres (thus neglecting the effects at the fibre ends), and Fig. 5.3a shows unit length of the composite and the associated expansions. The axial displacements per unit length (i.e. strains) of this composite are

$$\varepsilon_{cz} = \varepsilon_{fz}^T + \varepsilon_{fz} = \varepsilon_{mz}^T + \varepsilon_{mz} \tag{5.1}$$

where ε_{fz}^T and ε_{mz}^T are the displacements of the fibre and matrix if free to expand thermally, and ε_{fz} and ε_{mz} are the elastic strains of the fibre and matrix resulting from their need to expand the same amount when stuck together in the composite.

The corresponding radial displacements are shown in Fig. 5.3b. We assume that the fibre and matrix do not separate. Thus

$$u_c = u_f^T + r\varepsilon_{fr} = u_m^T + u_m \tag{5.2}$$

We also assume uniform thermal expansion, so $u_f^T = r\varepsilon_{fr}^T$ and $u_m^T = r\varepsilon_{mr}^T$. In addition ε_{fr} is the fibre elastic strain and u_m is the elastic matrix radial displacement, given in standard elasticity texts as

$$u_m = Pr \{ v_m + [1 + V_f]/ V_m \}/E_m \tag{5.3}$$

where P is the internal pressure in the tube of matrix, resulting from the force exerted by the fibre expansion. For this "composite" $V_f = r^2/R_m^2$ so we have replaced the r^2 and R_m^2 terms in the standard expression by V_f and V_m. In our case we must superpose the strain $-v_m \sigma_{mz}/E_m$ because of the axial stresses. Thus

$$u_m /r = \{P /E_m \}\{ v_m + [1 + V_f]/ V_m \} - v_m \sigma_{mz}/E_m \tag{5.4}$$

The stresses are

$$\sigma_{fr} = \sigma_{f\theta} = \sigma_{mr} = -P \tag{5.5}$$

and

$$\sigma_{m\theta} = P (1 + V_f)/ V_m \tag{5.6}$$

Equilibrium in the axial direction in the absence of externally applied stresses requires that

$$V_f \sigma_{fz} + V_m \sigma_{mz} = 0 \tag{5.7}$$

Now we use two of the three general stress-strain equations, Eqs. (1.8 & 1.10) adapted for polar co-ordinates for the fibre, thus:

$$\varepsilon_{fr} = (\sigma_{fr} + v_f[\sigma_{f\theta} + \sigma_{fz}])/ E_f \tag{5.8}$$

$$\varepsilon_{fz} = (\sigma_{fz} + v_f[\sigma_{f\theta} + \sigma_{fr}])/E_f \tag{5.9}$$

and for the matrix likewise, Eq. (1.10) becomes:

$$\varepsilon_{mz} = (\sigma_{mz} + v_m[\sigma_{m\theta} + \sigma_{mr}])/E_m \tag{5.10}$$

Now consider the radial strains. Writing $\Delta\varepsilon_r$ for $\varepsilon_{fr}^T - \varepsilon_{mr}^T$, Eq. (5.2) gives

$$\Delta\varepsilon_r = u_m/r - \varepsilon_{fr} \tag{5.11}$$

So substituting for u_m/r using Eq. (5.1) and ε_{fr} from Eq. (5.8) with $\sigma_{f\theta}$ and σ_{fr} given by Eq. (5.5) yields

$$\Delta\varepsilon_r = \{P/E_m\}\{v_m + [1+V_f]/V_m\} - v_m\sigma_{mz}/E_m + \{P/E_f\}\{1-v_f\} + v_f\sigma_{fz}/E_f \tag{5.12}$$

Rationalizing this, using Eq. (5.7) for σ_{mz}, gives

$$V_m E_m E_f \Delta\varepsilon_r = E_\lambda P + E_v \sigma_{fz} \tag{5.13}$$

where E_λ and E_v are abbreviated complex moduli given by

$$E_\lambda = E_1 + E_f + V_m\{v_m E_f - v_f E_m\} \tag{5.14}$$

$$E_v = v_m V_f E_f + v_f V_m E_m \tag{5.15}$$

with $E_1 = V_f E_f + V_m E_m$, i.e. the Rule of Mixtures, Eq. (3.3)

For the axial strain we write $\Delta\varepsilon_z = \varepsilon_{fz}^T - \varepsilon_{mz}^T$, so that Eq. (5.1) gives

$$\Delta\varepsilon_z = \varepsilon_{mz} - \varepsilon_{fz} \tag{5.16}$$

which using Eqs. (5.5, 5.6, 5.9 & 5.10) becomes

$$\Delta\varepsilon_z = v_m\{P/E_m\}\{1 - [1+V_f]/V_m\} + \sigma_{mz}/E_m - \{1/E_f\}\{\sigma_{fz} + 2v_f P\} \tag{5.17}$$

Again rationalizing and using Eq. (5.7),

$$V_m E_m E_f \Delta\varepsilon_z = -\{2E_v P + E_1\sigma_{fz}\} \tag{5.18}$$

Finally eliminating P between Eqs. (5.13 & 5.18) gives

$$\sigma_{mz} = V_f E_f E_m \{2E_v \Delta\varepsilon_r + E_\lambda \Delta\varepsilon_z\}/\{E_1 E_\lambda - 2 E_v^2\} \tag{5.19}$$

where we have exchanged σ_{mz} for σ_{fz} using Eq. (5.7). Also eliminating σ_{fz} between Eqs. (5.13 & 5.18) gives

$$P = V_f E_f E_m \{E_1\Delta\varepsilon_r + E_v \Delta\varepsilon_z\}/\{E_1 E_\lambda - 2 E_v^2\} \tag{5.20}$$

σ_{mz} and P can be quite significant in carbon and glass fibre laminates, but owing to the radial CTE of Kevlar, only σ_{mz} is significant in Kevlar composites.

For $\Delta\varepsilon_r$ and $\Delta\varepsilon_z$ we need a knowledge of the variation of the CTE's with temperature. Thus,

$$\Delta\varepsilon_r = \int_{T_i}^{T_f} (\alpha_{fr} - \alpha_{mr})dT \qquad (5.21)$$

and

$$\Delta\varepsilon_z = \int_{T_i}^{T_f} (\alpha_{fz} - \alpha_{mx})dT \qquad (5.22)$$

where T_i and T_f are the initial and final temperatures. We will leave the application of these equations to the problems at the end of this chapter.

The stresses, σ_W, responsible for the warping in Fig. 5.1 can be calculated from the radius of curvature, R_W.

$$\sigma_W = \frac{h}{24R_W}(E_1 + E_2) \qquad (5.23)$$

for a total laminate thickness of h. For a strain difference between layers, $\Delta\varepsilon$, due to cure or differential thermal shrinkage:

$$R_W = \frac{h}{2\Delta\varepsilon}\left[1 + \frac{1}{12}(E_1 + E_2)\left(\frac{1}{E_1} + \frac{1}{E_2}\right)\right] \qquad (5.24)$$

The derivation of Eqs. 5.23 and 5.24 and approximate expressions for $\Delta\varepsilon$ will be taken up as problems at the end of this chapter.

5.1.2. Creep and stress relaxation

Materials under constant stress can gradually extend by plastic deformation. The stress to cause this *creep* can be considerably less than the tensile strength. With metals the process is insignificant at temperatures less than about half the absolute melting temperature. Thus the creep of the metals normally used for bridges and aircraft (steel and aluminium alloys), for example, is negligible at room temperature. Polymers, however, can creep significantly at room temperature.

Fig. 5.4 shows typical creep curves obtained with steel and an aluminium alloy. It may be seen that the curves have three regions. The first, primary creep region, has a relatively high creep rate. This decreases with time until a constant creep rate is achieved in the secondary creep region. Eventually the specimen has deformed to such an extent that necking starts, and the creep rate increases once again. This is the tertiary creep region, and it generally continues for a short period only, whereupon the specimen fails, at the *creep rupture* point.

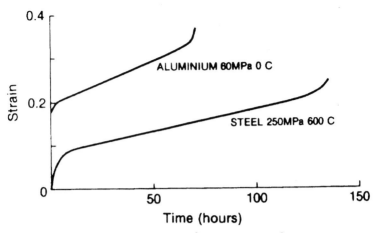

Figure 5.4. Creep curves for aluminium at $0°C$ and steel at $600°C$, at stresses indicated.

Polymers creep at room temperature at stresses far below their ultimate strengths: see Fig. 5.5. The addition of ceramic fibres reduces the creep, and when stressed in a fibre direction, carbon- and glass-containing laminates have negligible creep. However, creep can still occur when the stress is not in the fibre direction, and can be substantial for quite small angles between the stress axis and the fibre direction.

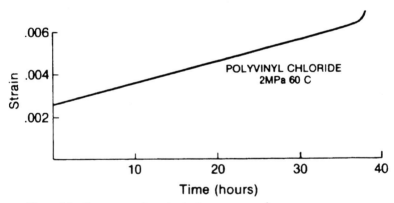

Figure 5.5. Creep curve for polyvinyl chloride at $60°C$ under a stress of 2MPa.

When biaxial stresses are present, laminates with fibres in the appropriate directions can keep creep rates down to a very low level. There is some residual creep, however, because the polymer between the lamellae can creep to a small extent.

In the case of glass fibre reinforced polymers, creep rupture occurs at stresses somewhat below the short term ultimate tensile strength, even though no creep strain is observed. In the region near the breaking stress, a 0.3% increase in stress decreases the life by a factor of ten. Also see Fig 4.15.

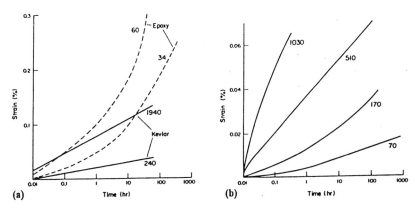

Figure 5.6. Room temperature creep curves for (a) Kevlar fibres and an epoxy resin, and (b) Kevlar-epoxy. The applied stresses (MPa) are marked on the curves. (After R. H Erikson, Composites, Vol. 7, 1976, pp. 189–194.)

In contrast with ceramic fibres, Kevlar (and other polymer fibres) do creep significantly, even at room temperature. Fig. 5.6a shows the creep of Kevlar fibres at two stresses at room temperature. Because of this, unidirectional Kevlar laminates creep at room temperature when stressed in the fibre direction. The creep curves of such a laminate, at 50% fibre volume fraction, have regions of primary creep and secondary creep. Results obtained, plotted as creep strain vs. log [time], are shown in Fig. 5.6b. These curves can be accounted for in terms of the creep of the matrix and fibres. The matrix controls the early stages of the creep process, while the fibres control the later stages. In addition, low stress creep depends on matrix creep, while creep at high stresses is governed by the fibre creep properties.

Stress relaxation is a natural consequence of creep. If a material is subjected to a constant displacement, the stress involved slowly relaxes. So when you fasten plastics together with screws, they loosen after a time. This can have serious consequences.

5.2. Aqueous environments

Water is absorbed by the polymer matrix, by some fibres used for reinforcement, and preferentially on some fibre-matrix interfaces.

5.2.1. The polymer matrix

Absorption of water softens and expands polymers to varying degrees, depending on their chemical nature. For instance, epoxies immersed in water absorb 3–4% by weight, depending on the temperature; see Fig. 5.7. This figure also shows the effect of varying amounts of humidity. PEEK only absorbs about 0.5%, Fig. 5.8.

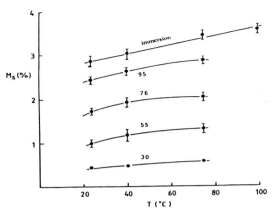

Figure 5.7. Temperature variation of water content at equilibrium of an epoxy resin for immersion and exposure at relative humidities indicated on the curves. (After M. Woo and M. R. Piggott, J. Comp. Tech. Res. , Vol. 9, 1987, pp.101-7.)

The rate of water uptake is governed by the *diffusion coefficient*, *D*, itself dependent on absolute temperature *T* (in degrees Kelvin), and governed by the *Arrhenius Equation*:

$$D = D_0 \exp[- \mathcal{E}_a / \mathbf{R} \, T] \tag{5.25}$$

where D_0 is a constant, $\mathcal{E}_a$ is the activation energy and $\mathbf{R}$ is the gas constant. The diffusion is normally governed by Fick's Law, and weight gain in a water environment has an initial linear region when it is plotted vs. $\sqrt{time}$, as shown in Fig. 5.8.

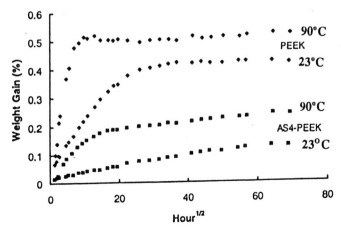

Figure 5.8. Water absorption of PEEK at 23°C and 90°C (upper curves) and absorption by a unidirectional carbon-PEEK laminate (lower curves). (After A. Zhang and M. R. Piggott, J Thermoplastic Comp. Vol. 30, 2000, pp. 162-72.)

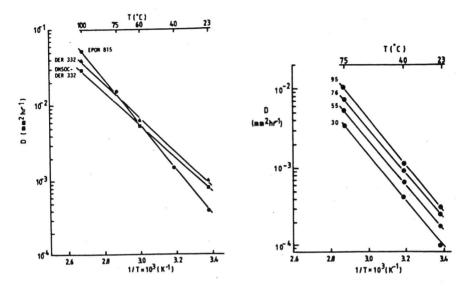

Figure 5.9. Arrhenius (1/T) plots for water diffusion coefficient, D. On the left is shown the water diffusion of three epoxies when immersed, and on the right, exposure of EPON 815 to various humidities, as indicated.

If the slope of this initial region is β, then the diffusion coefficient is given by

$$D = \pi h^2 \beta^2 / 16 M_s^2 \qquad (5.26)$$

for a sheet of thickness h and a saturated (or equilibrium) water content M_s. (The other dimensions of the sheet must be $>> h$). D obtained in this way, plotted against $1/T$, is shown in Fig. 5.9. On the left is shown the water diffusion of three epoxies when immersed, and on the right, exposure of one of these epoxies to various humidities.

Table 5.1. Typical values for water adsorption, M_s, and diffusion constant, D, of polymers

Polymer	Type	Environment	M_s (%)	D ($\mu m^2 s^{-1}$)
Nylon 66	TP[1]	90% RH, 20°C	4.8	3.6
PEEK	TP	Immersion, 20°C	0.42	0.98
Polyester	TS[2]	Immersion, 60°C	2.1	4.3
Epoxy	TS	Immersion, 20°C	3.0	0.12
Polyimide	TS	Immersion, 60°C	1.4[3]	0.57[3]

Notes: [1]thermoplastic, [2]thermoset, [3]estimated from composite results, assuming $V_f = 0.6$ and $D_c/D_p = 0.44$ (see Fig. 5.15 for $\sqrt{V_f} = 0.77$)

5.2.2. The fibres and the interface

Polymer fibres like Kevlar absorb relatively large amounts of water. Fig. 5.10. shows that reinforcing epoxy resin with Kevlar increases the water absorption, whereas the use of carbon or glass fibres reduces it.

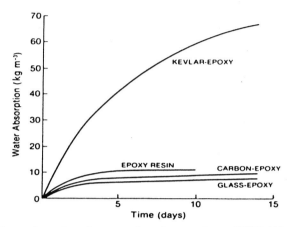

Figure 5.10. Moisture absorption of epoxy resin and composites at 100°C. (After D. C. Phillips, J. M. Scott and N. Buckley, Proc. ICCM2, 1978, pp. 1544-9.)

Ceramic fibres like glass and carbon do not absorb significant amounts of water, but they absorb some on the sized surface. (The *sizing* is a coating on the individual fibres; commercial ceramic fibres are normally coated to protect their surfaces from inadvertent damage, lubricate them so that they go easily through guides, and to promote adhesion with the resin matrix, see pp. 89-91.) Surface absorption is most easily demonstrated by exposing the fibres to high humidities; see Fig 5.11.

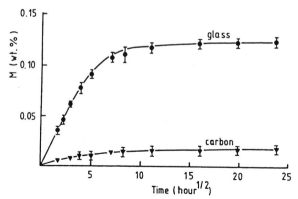

Figure 5.11. Water absorption of glass fibre rovings and carbon fibre tows at 98% RH and 23°C. (After M. Woo and M. R. Piggott, J. Comp. Tech. Res., Vol. 9, 1987, pp. 162-6.)

The fibre-polymer interface can be weakened by exposure of the composite to water. Manufacturers strive to make this weakening as limited as possible, and our tests have shown the bond can be very durable indeed. Fig. 5.12 shows some results of transverse test on commercial moldings. Subsequent examination of the fracture surfaces in the scanning electron microscope revealed that the fibres still retained thin films of polymer adhering to their surfaces. This shows that any weakening that did occur was probably restricted to the polymer matrix itself.

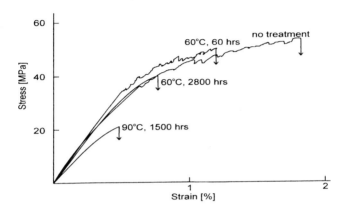

Figure 5.12. Stress-strain curves for transverse tests on unidirectional carbon-epoxy immersed in water. (After M. Chan and M. R. Piggott, Composites Interfaces, Vol. 6, 1999, pp. 543-556.)

(Perhaps the most sensitive test to detect the effect of water (and other fluids) on adhesion is the peel test (see p. 36, Fig 3.6). The paper referred to in the caption to Fig. 5.12 has some results on this. They show a gradual diminution in peel force with water exposure time.)

Preferential diffusion along the fibre-polymer interface has been demonstrated, at least for glass-epoxy; see Fig. 5.13. On the left is the set-up used, and on the right is a typical tritium exposure, showing black dots corresponding to diffusion along two glass fibres. The diffusion rate was determined by a series of exposures at ever increasing intervals.

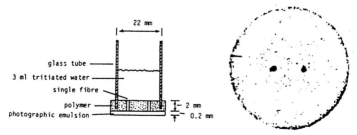

Figure 5.13. Tritiated water was used for this demonstration of preferential diffusion along the fibre interface. (After P. S. Chua, S. R. Dai and M. R Piggott, J. Mater. Sci., Vol. 27, 1992, pp. 919-24.)

5.2.3. The fibre composite

As shown in Figs. 5.8 and 5.10, the reinforcement of polymers, using ceramic fibres, decreases the saturated water content and D. The relative water content of a composite, V_{wc}, at saturation, normally follows a Rule of Mixtures type expression, thus

$$V_{wc} = V_f V_{wf} + V_m V_{wp} \tag{5.27}$$

where V_{wf} is the relative amount of water in the fibres and V_{wp} is the relative amount of water in the polymer matrix. Since ceramic fibres do not absorb significant amounts of water $V_{wf} \triangleq 0$, so the greater the fibre volume fraction, the smaller is V_{wc}. Glass-epoxy obeys Eq. (5.27), as can be seen in Fig. 5.14. Carbon-epoxy also has been shown to obey Eq. (5.27).

D is decreased as an approximately linear function of the square root of the volume fraction of fibres. This is because the fibres block the passage of water, and the relative size of the matrix passages is dependant on $\sqrt{V_f}$. For instance, the smallest gap between fibres in a regular hexagonal array is $R\left\{ 1 - \sqrt{2\sqrt{3}V_f / \pi} \right\}$ where R is the centre-to-centre fibre spacing. Thus the proportion which is clear of impediment is $\left\{ 1 - \sqrt{2\sqrt{3}V_f / \pi} \right\}$. For $V_f = 0.70$ this comes to 0.121.

So we can propose a lower limit to the diffusion coefficient, D_c, of a composite with impervious fibres as

$$D_c / D_p = 1 - \sqrt{2\sqrt{3}V_f / \pi} \tag{5.28}$$

where D_p is the diffusion coefficient of the polymer. Looking at actual results, Fig. 5.15, shows that this is much too low.

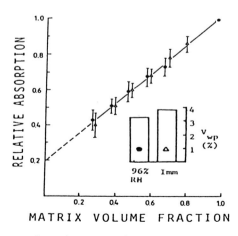

Figure 5.14. Relative water absorption at saturation. Immersion results (open triangles) and 96% RH results (filled circles) are shown. Inset is the weight % water absorbed by the resin alone.
(After M. Woo and M. R. Piggott, J. Comp. Tech. Res., Vol. 9, 1987, pp. 162-6.)

A full analysis, taking into account the interphasial material between the fibres and polymer, seemed to give a better fit, but an alternative model with voids, having rapid diffusion, given by $D_v = 15D_p$, seemed to fit the experimental points more closely.

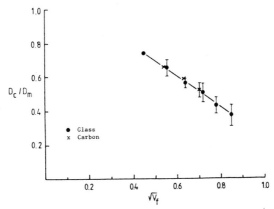

Figure 5.15. Glass-epoxy and carbon-epoxy diffusion results plotted vs. $\sqrt{V_f}$. (After M. Woo and M. R. Piggott, J. Comp. Tech. Res., Vol. 9, 1987, pp. 162-6.)

5.3. Other fluids

The effects of other fluids do not seem to have been as extensively investigated. Glass-polyesters (and to a lesser extent, glass-epoxies) are subject to attack by strong acids and alkalis. Carbon reinforced polymers are less susceptible. Carbon-PEEK and carbon-polyimide are very resistant. Aviation fluids affect some polymer matrices. Again, effects on epoxies are slight, more serious with polyester, and negligible with PEEK and polyimide.

Further reading

Anon (Editors of Modern Plastics Encyclopedia) *Guide to Plastics*, pp. 96-99, 1987.

C. E. Harris and T. S. Gates (Editors), *High Temperature and Environmental Effects on Polymeric Composites*, ASTM STP 1174, 1993.

T. S. Gates and A.H. Zurick (Editors), *High Temperature and Environmental Effects on Polymeric Composites*, ASTM STP 1302, 1997.

Problems

You are recommended to solve the following problems in the order given. Data needed will be found in tables in Chapter 2 and other chapters of the book.
Where a range of values is given, take the mean value; assume that the CTE's don't vary significantly with temperature.

(Note: Some of these problems, at least in the same outline form, have already appeared LBFC1 and LBFC2.)

5.1.　Imagine that the specimen shown in Fig. 5.1 is a [0,90] laminate (actually it is [0,±45,0] woven Kevlar-epoxy). Estimate the radius of the front edge of the coupon, and from it estimate what the thermal stress would be if it were indeed [0,90]. Assume each lamina is 1.0mm thick.

5.2.　Let $MR_W = EI$ on each half of a two ply laminate, the plies having moduli E_1 and E_2 and the same thickness $h/2$. The moment, M, is developed through the action of a mean compressive stress on one side, and a mean tensile stress on the other, both of magnitude P. The moment of area, I, is equal to $bh^3/12$, where b is the beam width and h its thickness. Hence develop Eq. (5.23).

5.3.　Imagine that the fibres in a laminate are all lumped together, in a way analogous to that shown in Fig. 3.3 and are virtually incompressible compared with the resin . Use a parallel matrix/fibre model to develop an approximate expression for α_1, the CTE of the $0°$ lamina and a series model for α_2 , the CTE of the $90°$ lamina. Hence develop an expression for $\alpha_2 - \alpha_1$.

5.4.　Develop Eq. (5.24) from the condition that the strain at the interface between the two layers shown in Fig. 5.3 must be the same.

5.5.　Calculate the radius of curvature at room temperature of a [0,90] E-glass-epoxy laminate, cured at 190°C. The 0.30mm thick laminae, have been made with a 31% volume fraction of epoxy, modulus 3.6GPa, and a CTE of 55 MK^{-1}. You may assume for now that $\Delta\varepsilon \triangleq V_m (\alpha_m - \alpha_f)\Delta T$, where the α's are the thermal expansion coefficients (with glass fibres the radial and axial CTE's may be assumed to be equal). ΔT is the difference between the cure temperature and room temperature (20°C). [Hint: calculate the moduli from the Rule of Mixtures and Eq. (3.11)].

5.6.　A [0,90] stiff carbon-epoxy laminate, with properties given in Table 2.8, had a radius of curvature after cure of 125mm. Each layer was 0.102mm thick. Estimate the curing stress and strain and the curing temperature assuming $V_f = 0.70$.

5.7.　Calculate the temperature at which transverse cracks would be likely to appear in a [0,90]$_s$ laminate made from the carbon-epoxy in the previous question, assuming that it was cured at 200°C. Neglect any thermal contraction in the $0°$ layers, and assume that they are rigid.

5.8.   A [0,90] S-glass-epoxy laminate, with properties given in Table 2.8, had a radius of curvature after cure of 105mm. Each layer was 0.122mm thick. Estimate the curing stress and strain and the curing temperature. For the epoxy E_m = 3.63GPa and the CTE is 64 MK^{-1}. (Hint: first estimate V_f and then E_2.)

5.9.   What would be the temperature in q. 5.7 if the $0°$ layer were not rigid and contracted thermally under the decreased temperature.

5.10.   What difference, if any, would it make to the previous problem, to superpose the microstresses that exist in the $90°$ layers? Assume v_f = 0.35, E_m = 3.91GPa and σ_{mu} = σ_{2u}.

5.11.   Estimate the axial shrinkage from the microstress equations, and compare it with α_1 for a $[0]_8$ laminate made from the same S-glass-epoxy material as in q. 5.8, assuming the Poisson's ratios of S-glass and E-glass are the same.

5.12.   Estimate the axial shrinkage from the microstress equations as in the above question, and compare it with α_1 for the composite in q. 5.10.

5.13.   How much extra strength, when tested at room temperature (20°C), might be contributed by σ_{fz}, in a $[0]_8$ laminate made from stiff-carbon-polyimide cured at 300°C. Would the polymer crack before the fibres break? The fibres constitute 65% of the volume.

5.14.   Estimate the diffusion coefficient at 23°C and at 90°C, and hence estimate the activation energy for diffusion from Fig. 5.8, given that the sample thickness was 3.0mm.

5.15.   The laminate in q. 5.6 absorbed water at a rate proportional to $\sqrt{time}$ until 60 seconds had elapsed, by which time it had absorbed 0.28% of its own weight of water. Estimate the diffusion coefficient, and for an activation energy 48 kJmol^{-1}, estimate the temperature at which the absorption took place. (Hint: use data from Table 5.1 and use Fig. 5.15 for D_c / D_p)

5.16   Assume M_{s90} / M_{s23} is the same for polymer and composite in the results shown in Fig 5.8, and M_{s90} is given by the result for composite immersed at 90°C 400h. Hence estimate D_c / D_p for the 23°C results. (Hint: you need the value that you obtained in q. 5.14).

Selected answers

5.1   64MPa

5.3 $\alpha_2 - \alpha_1 \stackrel{\Omega}{=} V_f \, \alpha_{fr} + V_m \, \alpha_m - \alpha_{fz}$

5.5   203mm

5.7 $-138°C$

5.9 $-137°C$

5.11 $10.9\ MK^{-1}$, underestimates the shrinkage by about 22%.

5.13   24MPa, i.e. 1.6%

5.15 $1.19\mu m^{2}s^{-1}$, 77°C

Chapter 6. How to Make and How to Use Composites

Having now learned how to design a composite and take into account the pitfalls of the reinforced plastic, we need to review the methods that are used to make composite structures like aeroplane wings, boats, bridges, canoes, cars, chemical plant, trains, prostheses, sports equipment and the countless other things made from these versatile materials. In addition, we will give examples to illustrate the wide range of employment of load bearing composites.

6.1. Starting materials

The final products, be it an aircraft wing or moving parts for machinery used by textile manufacturers, are very diverse. So too are the forms of starting material. The simplest is probably reels of fibre, together with cans of thermosetting resins. We will start with the fibres.

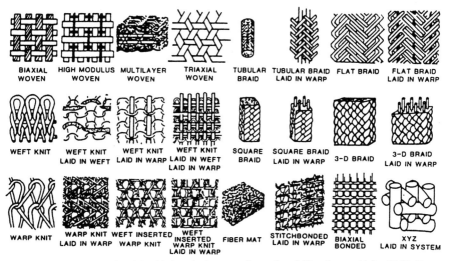

Figure 6.1. Woven, knitted, braided and other two dimensional fibre forms. (After F. K. Ko, Ceramic Bull., Vol. 68, 1989, pp. 401-14.)

6.1.1. The fibre surface

It is important to ensure that the polymer wets the fibres quickly and thoroughly, and that a good bond is achieved. Most fibres are coated with a thin film of polymer. However, other surface treatments are used, which are specific to the individual fibres, and have at least three roles.

1. *Surface protection.*   Ceramic fibres like glass and carbon are brittle and very easily damaged. These fibres are coated with a relatively soft polymeric material in

order to prevent them touching each other, or anything else, since even light contact can damage and weaken them. Starch is used for glass (it can be deposited from water) when fibres are sold to processors who want to do their own surface treating. Carbon is usually treated with epoxy resin, often not fully cured, and usually containing very little curing agent. Kevlar, not being brittle, is not usually surface treated.

2. *Lubrication.*   In the processing of fibres to produce the premixes, prepregs, preforms, filament windings, etc., described later, they have to be unspooled and taken to the site of operation through loops and eyes. These are often made of hard metals or ceramics. So to ease the passage of the fibres, the surface treatment of glass usually contains a lubricant.

3. *Adhesion promotion.*   Glass is coated with a material that bonds strongly to the fibres and has an organic "tail" that interacts beneficially with the polymer. The end that bonds with the glass is usually a siloxane, i.e. containing a $-SiO_3$ group. For thermosetting resins the organic tail contains a reactive group appropriate for epoxy, polyester or polyimide. For a thermoplastic the tail is a relatively long chain hydrocarbon which can penetrate a thermoplastic matrix during processing and form an *interpenetration bond*.

Carbon fibres have soft friable surface layers which must be removed. Oxidation is used for this, either with strong acids or oxidising agents, or by heating in the presence of oxygen. The resulting surface has good adhesion due to the presence of active groups such as $=CO$. The activity can be enhanced by plasma treatment, using high frequency high voltage electrical fields in a partial vacuum to produce the plasma.

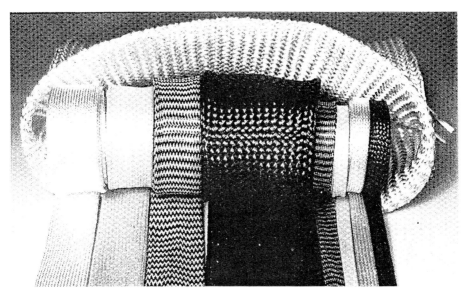

Figure 6.2. Examples of braided forms. (Courtesy of A & P Technology.)

Kevlar and polyethylene fibres, usually untreated, do not adhere well to the matrix polymer. However, adhesion can be improved by plasma treatment. Kevlar fibres should be dried before use.

6.1.2. Fibre forms

The fibres may be twisted (yarn) or untwisted (roving [glass] or tow [carbon]). Tapes, consisting of fibres aligned along the axis of the tape are used when the modulus has to be as large as possible. The fibres may also be woven, knitted or braided in a large variety of forms, Fig 6.1 and 6.2.

The woven forms are normally less flexible than knits and braids, but they can be flexible enough to be draped, without wrinkling, over surfaces with double curvature ("drapeable" weaves). Weaves impart higher stiffness to the composite than knits or braids. The modulus imparted by woven materials may be enhanced by putting more fibres in the critical direction. But that reduces the off-axis stiffness and the shear properties.

Preforms are shaped assemblies of fibres, often woven.. See fig. 6.3.

Figure 6.3. A carbon fibre preform (3D weave) used in the space program. (Courtesy of Hexcel.)

6.1.3. The polymers

As indicated in Table 2.8, the main polymers used for high performance composites are thermosets; they are much easier to mold than thermoplastics, such as PEEK. (Note, though, that the high temperature thermosetting polyimides are very difficult to mould). Sometimes thermoplastic polyimides are used for specially exacting environments, such as aggressive solvents and chemicals, and high temperatures.

Thermosets have to be mixed with curing agents prior to use, while thermoplastics must be melted. These characteristics determine the type of process used to manufacturer of reinforced plastic parts.

6.1.4. Premixes

It is often advantageous to combine the fibres with the resin before making the article. If a thermoset is used, it is combined with the fibres in a partly polymerized state, so that it can flow, when moulded, to take the desired shape. Once shaped, it is heated to complete the polymerization and make the shape permanent. Thermoplastics may also be combined with fibres prior to production. These are heated during moulding to melt the polymer. (Composites can also be moulded directly from mixtures of chopped fibres and powdered matrix or *sheet molding compounds* [SMC's], consisting of a sheet of polymer containing chopped fibres. These composites are generally too weak and too prone to creep to bear significant loads.)

The premixed materials have a number of important advantages. The quality of the product is reproducible, the fibre content can be very high, and the fibres can be uniformly distributed. Continuous fibre-resin mixtures are usually called *prepregs*, and can be in the form of tape or sheets, with woven or straight fibres. The resin used is either a thermoplastic or a partly cured thermoset. In the thermoplastic case the prepreg can be made with commingled polymer fibres and reinforcing fibres, or with powdered polymer. These are partly consolidated by heat and pressure. With thermosets the fibres pass through a bath of resin, and are partly cured in an oven, before being wound on spools, with release film between layers. Thermoset prepregs have to be stored at low temperatures, otherwise they harden prematurely. Glass, carbon, and Kevlar prepregs are widely available.

6.2. Manufacturing methods

A large number of different methods are used for the manufacture of the reinforced polymer product. The method chosen in any particular instance depends on the type and size of article being produced, the number of identical articles to be made, and the strength, stiffness, and other properties required of the material. In this section some of the many moulding methods used for the manufacture of reinforced plastic articles will be described.

This description is far from exhaustive; many specialized companies use many different approaches with these highly amenable materials. The methods may be divided into two categories, those that require significant pressures and those that do not.

6.2.1. Low pressure methods

1. ***Hand lay-up.***   The simplest method of making a composite is to lay the fibres onto a mould by hand, brush the resin on, and allow it to cure. This requires very little capital investment, and though labour-intensive, is still much used because of its great versatility.

The fibres can be in the form of random mat or cloth, and very large structures can easily be built if room temperature curing resins (e.g. epoxy or polyester) are used. The resin can be sprayed rather than brushed on. The moulds do not need to be particularly strong, and can be made, for example, with balsa wood, and covered with plaster. More permanent moulds can be made with reinforced plastics.

Fig. 6.4 illustrates this process. First the mould is coated with a release agent to ensure that the part can be removed. Next the gel coat is applied. This is usually a pigmented resin layer to give good appearance, and may be sprayed on. When this is tacky, the fibre mat or cloth is laid on manually, and more resin is applied by pouring, brushing, or spraying. Next a roller or a squeegee is used to ensure thorough impregnation and wetting of the fibres. (The removal of trapped air is particularly important.) Further layers are then added in the same way until the required thickness is obtained. The curing of the moulded part may be accelerated by heating.

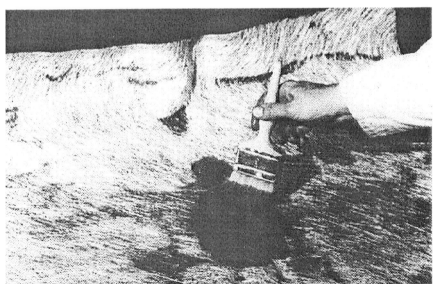

Figure 6.4. Resin is applied by brush in the hand lay-up process. (Courtesy Fibreglass, UK.)

2. **Spray-up.**   In this method the fibres and resin are sprayed together onto the mould. The layers deposited are densified with rollers or squeegees as for hand lay-up. Gel coats are often used for good surface finish. Spray-up can be used to mould more complex shapes than hand lay-up. However, since the fibres have to be chopped in the spray gun, the composite produced cannot be as strong or stiff as hand laid-up composites with cloth or other continuous fibre forms.

Polyester or other themosetting resins are usually used. Very large parts are cured at room temperature, but with smaller parts the curing may be accelerated by bag moulding (Section 6.2.2). The moulds used do not need to be strong, and can be made with wood and plaster, or fibreglass. Preforms (see Fig 6.3) are often used for spray-up

3. **Filament winding.**   This   process   consists   of   winding   continuous filaments over a suitably shaped mandrel. The filaments, as rovings (fibreglass) or tow (carbon) are impregnated with resin just before they go onto the mandrel for the wet winding process (Fig. 6.5). There is also a dry winding process; in this, prepreg tapes are used. Epoxy resins are often used for the matrix, though polyester and vinyl ester and other resins may also be used. Methods for filament winding reinforced thermoplastics have been developed recently. These involve heating at the line of contact between the material (prepreg, or reinforcing fibres plus polymer fibres or polymer film) and whatever is already present on the mandrel. A similar approach for thermosets uses electron beam curing.

It is usual to rotate the mandrel, though a stationary mandrel is sometimes used. Two winding methods are available; polar (or planar) winding, in which each layer of fibres is wound without spaces or cross-overs, and helical winding, in which both spaces and cross-overs occur. In both winding processes the fibres are laid onto the mandrel in a helical pattern, and the helix angle is chosen to suit the application. The arrangements of the fibres at the ends are most important for pressure vessels, and poorly designed fibre patterns can lead to early failure at the ends.

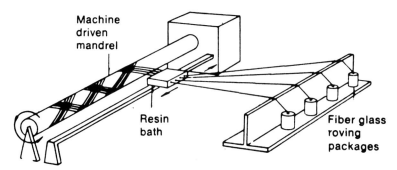

Figure 6.5. Filament Winding. (Courtesy PPG Industries.)

The construction of the mandrel requires considerable skill. It must not collapse under the large pressure resulting from the fibre winding tension, and must be easily removed when the process is complete. Segmented metal forms (usually steel) are most commonly used, and they may be faced with plaster. This method produces very strong composites, and very large cylindrical and spherical vessels can be built.

4. *Fibre placement.* Filament winding can only be used to place fibres on convex surfaces. On flat surfaces the fibres and resin are not properly consolidated, and concave surfaces are bridged by the fibres. The fibre placement method eliminates these problems. It uses a robotic head which can move in three dimensions. For large structures the robotic assembly can be moved on rails along the length of the specimen; see Fig. 6.6. The mandrel on which the material is laid down may rotate or be stationary. The head (see the inset) can be heated to partly cure the resin as it places the material, or melt the resin, as required for a thermoset or thermoplastic. Also, the electron beam curing method may be used.

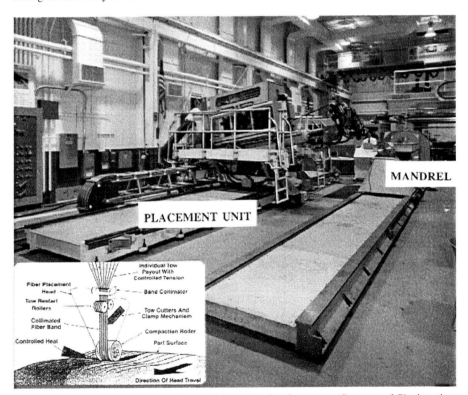

Figure 6.6. Fibre placement machine with inset, details of process. (Courtesy of Cincinnati Machine.)

5. *Pultrusion.* This is a method which is used to make very strong aligned fibre composites. The fibres are impregnated with resin and pulled through a die shaped to produce the desired cross-section in the product. The method is suitable for use with thermosets. The die is heated to promote setting of the resin after it has impregnated

the fibres; see Fig. 6.7. Further heat may be applied either before the composite enters or after it leaves the mould. When the material has hardened it is cut into suitable lengths.

Many types of sections are available from the pultrusion process: see Fig. 6.8. Furthermore, woven materials can also be pultruded, so that the strength normal to the axis can be increased. This is particularly useful for structures such as I beams. Finally, very large pultrusions for such structures as bridge decks can now be manufactured.

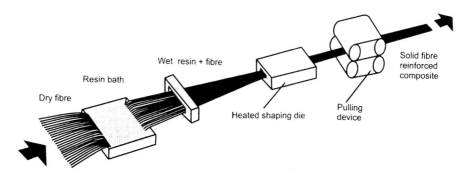

Figure 6.7. Continuous pultrusion.

Recently, special thermoplastics have been developed for pultrusion. See Section 9.1. in LBFC2 for the description of these and other polymers.

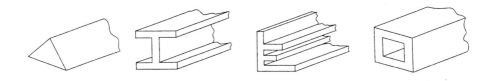

Figure 6.8 Examples of small sections which may be pultruded.

6.2.2. High pressure methods

1. *Tube rolling.*   The principle of the tube rolling method is illustrated in Fig. 6.9. The method is used to produce finite lengths of tube, usually of small diameter, for such uses as golf club handles, billiard cues, fishing rods, etc. Diameters range from a few mm up to about 100mm. Prepregs are most suitable for rolling, and woven fibre prepregs are most common.

It is usual to cut the material out into the shape required to make the complete roll, and it is important to have the fibres appropriately oriented. This is called convolute rolling. Alternatively, prepregs can be spirally rolled, a process that can be mechanized. Both methods may be used to produce tapered tubes. Special rolling tables are available, together with automatic prepreg cutting machines to cut the prepreg into the special shapes required.

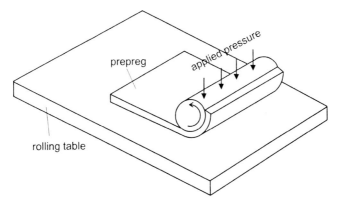

Figure 6.9. Principle of tube rolling.

2. *Cavity moulding.*   Injection moulding requires very large pressures and a mould which has the appropriately shaped cavity for the piece to be molded. The method is widely used for thermoplastic polymers without reinforcement, but low and high performance fibre composites are also made this way. (Low performance fibre composites typically contain chopped glass fibres which have been further degraded by extrusion, so that their lengths are 1 mm or less.)

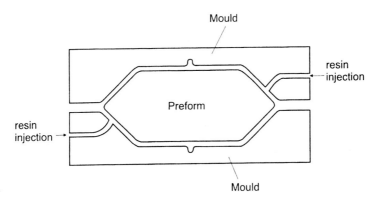

Figure 6.10. Principle of resin transfer moulding.

A similar method can be used for resin transfer moulding (RTM), which normally uses a thermoset, and hence can be carried out at low pressures, see Fig. 6.10. In this case a "preform" is made in a separate operation. This preform consists of fibres with a small trace of resin, which has been shaped, as required for the final structure. The spray-up method is often used for this, but preforms can also consist of three dimensionally woven structures made with continuous fibres (see Fig. 6.3).

The preform is inserted into the mould cavity, and the resin-hardener mixture is injected therein; see Fig. 6.8. Heating hardens the resin to complete the process. To ensure good impregnation, the air may be evacuated from the mould before resin transfer. This process is also referred to as structural RRIM (i.e. reinforced reaction injection moulding). The structural RRIM is also sometimes referred to as SRMM. (We deplore the alphabet soup which develops when fibres are added. Some acronyms are listed in Appendix C.)

3. *Bag moulding.* This method may also be used for making large parts. There are three different ways in which the moulding may be done, as shown in Fig. 6.11. The pressure bag system is relatively expensive since the combined mould-pressure vessel can only be used for one shape. The vacuum bag and the autoclave are very versatile, and relatively economically.

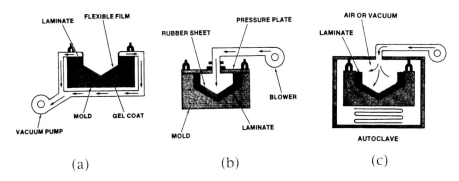

Figure 6.11. Bag moulding methods. (a) vacuum, (b) pressure, (c) autoclave. (Courtesy of Fiberglas Canada, now Owens Corning.)

The aircraft industry is a major user of the autoclave method, and employs very large autoclaves to manufacture wing parts, etc. Prepregs which are precut on automatic cutting tables are normally used. Moulds are made from steel, nickel, aluminium, graphite, reinforced plastics, or plaster. Laying up the part is a major operation, since it involves first, cutting the prepreg into the appropriately shaped pieces, then laying them up on the mould, with layers in the required orientations. Parts, such as honeycomb layers are added, together possibly with fasteners or any other components which are required to be moulded in.

Beneath the laminate structure there is a release cloth to ensure that the laminate can easily be removed from the mould (usually called the "tool"). More release cloth goes on top, then bleeder layers (often random fibre mat) then a "breather" layer to facilitate air removal, typically made from a polyester fibre mat for mouldings at 120°C or less. Finally, there is the vacuum bag, and also around the edge, a sealant tape or plasticene-like material, together with vacuum fittings: see Fig. 6.12.

The whole assembly is evacuated, then rolled on a carriage into the autoclave which has vacuum lead-throughs, so that the vacuum can be maintained during cure.

Once all the parts are in the autoclave, pressure is applied, and the whole assembly inside is then cured. This may involve several heating steps with controlled rates of heating and cooling, usually with microprocessor control.

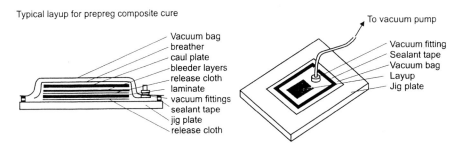

Typical layup for prepreg composite cure

Vacuum bag
breather
caul plate
bleeder layers
release cloth
laminate
vacuum fittings
sealant tape
jig plate
release cloth

To vacuum pump

Vacuum fitting
Sealant tape
Vacuum bag
Layup
Jig plate

Figure 6.12. Schematic drawing of the arrangement used for autoclave moulding of aircraft parts.

6.3. Joints

In a complicated structure it is usually necessary to divide the structure into more elementary parts, construct these, then join them together.

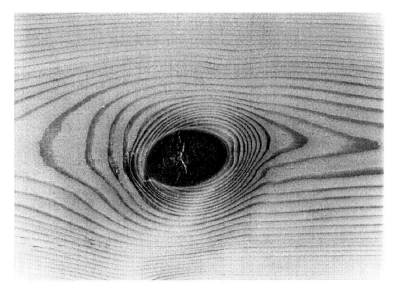

Figure 6.13. Nature's joint. Note how the grain changes around the joint, indicating a change in fibre orientation.

Composites present special problems for joining because of their extreme anisotropy. Even cross-ply laminates have problems because of their low out-of-plane properties. Nature has solved the problem (Fig. 6.13) by arranging fibre growth around the joint so that it has optimum properties. This approach is not usually available to designers of composites.

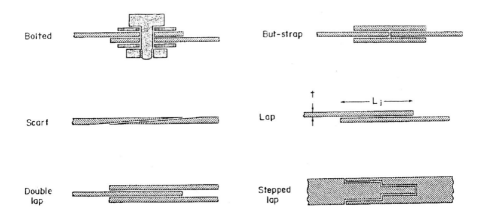

Figure 6.14. Types of joint suitable for joining laminates and sheets.

The most cost-effective methods of joining isotropic materials are usually either bolts or rivets. Rivets are seldom suitable for composites, but bolts are often cost effective in joining laminates to each other, or to sheets of other materials. Due to the low apparent shear strength of the composite the bolt clamping load should be spread over a large area by the use of washers, as shown in Fig. 6.14. The washers should be as large as possible, consistent with weight requirements.

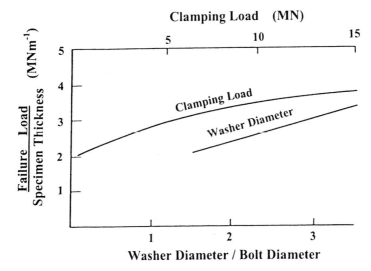

Figure 6.15. Effect of clamping load and washer diameter on the load-carrying capacity of a bolt in a glass fibre laminate. (After Stockdale, J.H. and Matthews, F.L., 1976, Composites 7, 34-38.)

Fig. 6.15 shows that the strength of the joint is a linear function of washer diameter. It is usually an advantage to have the clamping load high, but the maximum is limited by the material. Fig. 6.15 also shows that with glass-epoxy the joint strength is greatly increased by good clamping, though there is little to be gained by having the load greater than about 15MN.

Bolted joints should be well fitting for maximum effectiveness. They should not be too close to the edge of the sheet, but about five bolt diameters is usually sufficient. They should only be used with extreme caution for unidirectional fibre composites, though, since they can very easily shear out under load.

Glued joints are normally used for unidirectional fibre composites. Butt-strap, scarf, and lap joints of various types are the main forms used for unidirectional composites and laminates. Examples are shown in Fig. 6.14. In all these joints the length of the joint is calculated assuming that the adhesive has an apparent shear strength τ_{au}. Thus, for sheets of thickness t, and a joint with a length of L_j, the single lap joint can transmit a stress σ_{ju} where

$$\sigma_{ju} = L_j \tau_{au} / t \tag{6.1}$$

Eq. (6.1) only works, however, for thin laminates, as shown in Fig. 6.16. Stress concentrations occur in this type of joint, which can be reduced by chamfering the edges. Moreover, as we showed in section 4.1, shear failure is not normally observed with polymers. The stress concentrations become serious when laminates with more than four layers are joined in this way. Thus, the effective value of τ_{au} is reduced to a small fraction of its value for thin laminates. In this case the butt-strap, scarf, and single or double lapped joints should not be used. Instead the stepped lap should be employed.

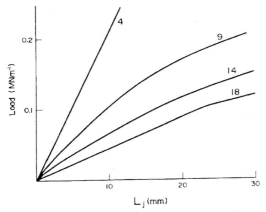

Figure 6.16. Load on each ply per unit width of joint as a function of lap length, L_j, for glued double-lapped glass-polymer/aluminium joints. The figures on the curves indicate the numbers of layers in the laminates. (After Grimes, G.C., and Greimann, L.F., 1974, Composite Materials 8, 1-35, Ed. Chamis, C.C., Academic Press.)

When a laminate has to be fixed to a massive support, a different type of joint is required. Some examples are shown in Fig. 6.17. The plain insert is the simplest and weakest. The divided root is much better, and the strength is greatly influenced by the number of divisions. The wrap around may be very promising, but is difficult to construct.

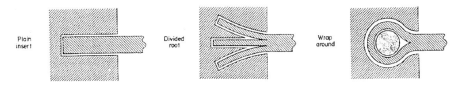

Figure 6.17. Joints to connect laminates with a massive support.

6.4. Applications

Reinforced polymers have been available to mankind since time immemorial, in the form of different kinds of wood. Wood can have excellent mechanical properties and a fighter aircraft, the Mosquito, made with it, performed well during the Second World War. Wood, however, is biodegradable, and the advent of synthetic fibre composites in the early 1940's opened up innumerable new possibilities. The use of glass-polyester for aircraft radomes was the starting point here, and glass fibre reinforced polymer was used successfully to make a small aeroplane in the 1940's. The small aeroplane use has continued, but the adoption of reinforced plastics was relatively limited to start with. Sports equipment, such as fishing rods, sailplanes and sailing dinghys, were early candidates, and wood was phased out quite quickly for sailplanes and small boats because glass-polyester ones performed better and were more durable.

New fibres with much higher modulus, such as the carbons and polyaramids, created further possibilities. Light materials which have better load carrying capacity than aluminium became available, so military aircraft started to make extensive use of reinforced plastics and they started to capture many new markets. Today we see a wide diversity of uses which are classified herein according to the industries that use them.

6.4.1. Aerospace structures

Initial use of reinforced polymers on commercial mid-sized aircraft followed the principle that failure must not risk loss of life. So fairings (which merely smooth the air flow) were among the first uses. Leading edges for the wings of small aircraft are also good places for composites, and de Havilland's Dash 8 used woven Kevlar-epoxy for those, after exhaustive tests with turkey carcases to ensure that they stood up well to bird strikes.

Floors of large aircraft are made with carbon-phenolic sandwiches with polymer honeycomb between the faces. These are routinely tested for resistance to damage caused by rolling trolleys. The matrix polymer used inside the aircraft – a phenolic – is one which produces minimal toxic fumes when on fire. Most of the other internal structures (inner walls, toilet cubicles, etc) are also made of reinforced plastics. But seat supports are still made with metal.

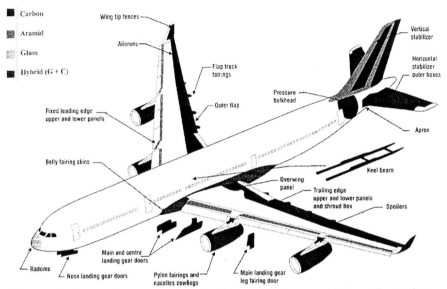

Figure 6.18. Composite components on the Airbus A320. (Courtesy of Hexcel Composites.)

Nowadays some specialized small aircraft have carbon-epoxy throughout their structures: a good example is the Voyageur, which made a non-stop flight round the world. Meanwhile, carbon reinforced plastics are just starting to be used for the stressed members of large civilian aircraft. Almost 20% of the dry weight of the Airbus 340 is composed of composites. Load bearing members include the keel beam made from carbon-epoxy, the vertical stabilizer, ailerons, spoilers, vanes and flaps from carbon-, glass- and aramid-epoxies, and the rudder and horizontal stabilizer from carbon- and glass-epoxies; see Fig. 6.18. The Boeing 777 vertical stabilizer is made from carbon-epoxy. As experience is gained, more will follow.

Aeroplane engines have also been made lighter and better with reinforced plastics; the Rolls Royce RB108 had glass fibre reinforced compressor rotor blades and casings in the 1950's. Nowadays these parts are made with carbon-epoxies, and for the parts experiencing higher temperatures carbon-polyimides are used. Reinforced plastics now comprise nearly one half the materials used in these engines.

Helicopters are now almost wholly made from reinforced polymers. Not only is the weight much reduced, but the use of reinforced plastics for the rotor blade and for the rotor head simplifies design, see Fig. 6.19, and endows these parts with much longer lives as compared with metals. Gliders also are now made from reinforced plastics.

In space applications composites play a major role. Rockets are filament wound carbon-epoxies, or carbon-polyimides for higher temperatures. Rocket motors too are made from carbon-epoxy: for the space shuttle, replacement of the metal reduced the motor weight by about 35% and used nearly 20 tonnes of the composite for each motor. The low CTE of the carbon fibre is a special bonus in some space applications. Thus the use of carbon-epoxy for the structure of the Hubble space telescope gives it greatly improved thermal stability compared with other candidate materials.

Figure 6.19. Reinforced polymer helicopter driving parts (a) section of the rotor blade molding. (Courtesy of Westland Helicopter Ltd.,) and (b) Rotor head. (Courtesy of Societe Nationale Industrielle Aerospatiale. Note that the design has been simplified and improved since this picture.)

In practice, aircraft structures almost universally use $0°$, $90°$, $\pm 45°$ arrangements in their laminates. This applies even to military aircraft. For example, a fighter aircraft used in Canada has, in one part of the wing, twenty eight $0°$ plies, six $90°$ plies and thirty four pairs of $+45°$ and $-45°$ plies. Other parts of the wing are constructed in a similar fashion. The stated reasons are as follows: 1) load paths are variable, and not always as expected; hence the four chosen orientations ensure safety; 2) data on such things as bearing capacity (at joints etc) are not available; 3) there is more material waste when cutting at other angles.

A more innovative approach to aircraft design is described in Section 3.3.2.

6.4.2. Marine structures

Glass reinforced polymers are well established in the boat-building industry, having been used since the late 1940's. Their characteristics of light weight and high strength, design flexibility, and low thermal conductivity are very advantageous in this application. The monolithic seamless construction minimizes assembly problems and leakage, while maintenance and repair costs are reduced. The most important advantage is their excellent resistance to the marine environment. Polyester resins are most commonly used for the matrix, being cheaper and easier to handle than epoxy resins. With glass-polyester laminates the loss in strength due to water in moderate climates is some 10-15%, increasing to a maximum of 20% in warm climates. The strength loss occurs during the first two months, and is due to the water penetrating the resin and acting as a plasticizer. If the fibres do not have the appropriate sizing the water is absorbed on the glass, greatly weakening the fibre-matrix interface and causing a large reduction in strength. Improperly finished or under-cured resins, or laminates with excessive voids, are also more susceptible to water degradation. Thus care is needed when fibreglass is used for construction.

Figure 6.20. America's Cup boat. (Courtesy of Hexcel.)

There is an increasing tendency to use epoxy resins. They have better resistance to moisture and weathering. Also, they are practically odorless when being molded, not having the troublesome styrene emission which accompanies the use of polyesters. Moreover, prepregs can be used, taking the chemistry out of the production process.

The glass employed is usually E-glass. The form of reinforcement used is random mats, spray-up, or woven cloths, according to the type of boat. A very large fraction of small boats are now made from fibreglass, and ± 60° weave has advantages for certain boat shapes. Such fabrics are available (e.g. Ahlstron Glass Fibre Oy) and can reduce weight by perhaps as much as 10% compared with the normal 0/90 weaves.

The use of fibreglass is being extended to ever larger boats using sandwich construction. A notable example is a 47 m minesweeper. Although costing more than steel, its advantages (being non-magnetic and more corrosion resistant) were worth the extra expense.

For boat sizes larger than 50m the use of fibreglass is presently uneconomic. These are made with steel hulls. Here, improved performance is obtained by using fibre composites for decking and internal structures. As with aircraft, phenolic resins are used as matrix in internal structures.

Kevlar and carbon are having a big effect on high performance boat construction. The light weight and extra stiffness are the main advantages here. Kevlar is mainly used in hybrids with glass or carbon because it is otherwise prone to failure by buckling, due to its weakness in compression. Boats, such as those used for the America's Cup (see Fig. 6.20) use a lot of carbon-epoxy for the hull, mast and other structures.

Reinforced plastics are starting to replace wood for shore structures such as piers and jetties. Again the reason is their superior durability.

6.4.3. Ground transport

Racing cars are nowadays made of carbon-epoxy. Fig. 6.21 shows the internal structure of such a car.

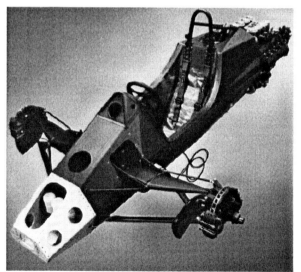

Figure 6.21. Internal structure of a racing car. (Courtesy of Hexcel.)

The major potential user is the private car, but this is still mostly made of steel at present because of the speed of manufacture and low cost. Specialty cars use more glass reinforced plastic – for doors, hoods, trunk lids, engine compartments etc. Sheet molding compounds are much used for this market. Sports cars have fibreglass bodies, but there is little use of higher performance fibres such as carbons or aramid. Government sponsored efforts in North America to reduce car weights have been side stepped by people buying vans and sport utility vehicles (SUV's), but the low volume (i.e. expensive) car market thrives because of the use of reinforced plastics.

Trucks make much use of reinforced plastics, but mainly for non load bearing parts. An important innovation here though is the fibreglass leaf spring. This is much lighter than the steel counterpart, has excellent fatigue resistance and fails in a safe, splitting mode, rather than fracturing across the thickness. A carbon-epoxy leaf spring has also been tested. This is so light that a spring for a lorry (or truck) can be replaced quite easily, since it weighs only about 7.4kg instead of the 60kg for a steel one. The fibreglass spring is widely used in some European cars.

Other commercial vehicles such as buses and street cars are similarly being made using larger amounts of fibreglass.

Figure 6.22 High speed tilting train uses reinforced plastics for bogies (driving wheel assemblies) and leaf springs as well as floors and side panels. (Courtesy of Adtranz.)

Freight containers are also being made from reinforced polymers rather than metals. Although the composite is initially more expensive than steel or aluminium, it has proved to be more durable, and hence cheaper in the long run. Glass-polyester foam sandwiches are particularly favoured for air freight, where their light weight is a great advantage. Chassis-less tanker trucks have evolved through the use of reinforced plastics. Weight saving has driven the use of composite materials for external and internal structures in railway engines and railway coaches. These include floor panels, and internal stiffening arches, for example on the German Regio Swinger high speed trains, which tilt when cornering: see Fig. 6.22. These also incorporate carbon- and glass-epoxy for leaf springs and bogie structures. Box cars make extensive use of reinforced plastics too.

Weight saving is also important in some military equipment. Tanks which can be carried in planes have hulls made with glass reinforced polymer.

In all transport applications, aeroplanes included, a major advantage is the ease with which aerodynamic shapes can be produced. Moreover, the molding process makes possible excellent finish which also reduces air drag.

6.4.4. Energy and energy storage

Aeroplane engines using reinforced plastic compressor blades have already been mentioned in Section 6.4.1. Another important area in energy conversion is windmills for electricity generation. Wind power is the world's fastest growing energy source, using wind driven turbines up to 66m in diameter and delivering 1.65MW of power, see Fig. 6.23. The preferred structure for the blades is woven glass-epoxy faced honeycomb. In both engines and windmills, the correct aerodynamic shape is crucial for efficiency, and the smooth finish produced by the molding process is a major benefit.

Figure 6.23. Modern windmills generating electricity in a wind farm in Germany. (Courtesy of Vestas Wind Systems.)

Gases used as fuel require pressure-resisting storage tanks. These are biaxially stressed in the ratio 2:1 so are commonly filament wound with a winding angle (relative to the tube axis) of $54.73°$; see Fig. 3.22. In the energy industry they are used for storage of natural gas, and when used to fuel buses and trucks, carbon fibres are used to reduce the dead weight. Moreover, pressure containers that permit more efficient packing are being developed, see Fig. 6.24.

Figure 6.24. Natural gas pressure storage tank designed for efficient use of space. (Courtesy of Thiokol Propulsion, a Division of Cordent Technologies.)

Oil production and storage uses filament wound reinforced plastics. Of particular interest here is the off shore oil rig. The risers which carry oil from the sea bed need to be corrosion and fatigue resistant. Much effort is going into the development of filament wound carbon-epoxy tubes to replace steel. They do not sag as much and hence lose rigidity. An additional benefit is that they can serve at the same time both as risers and tethers.

Electric transmission lines make much use of filament wound fibreglass poles, and insulators are also made with fibreglass. Apart from very much reduced weight, the insulator is resistant to vandalism and rifle fire, which is a major problem in the insulator industry.

Air and oxygen storage for firefighters and divers etc. require lightweight tanks which are impervious to the gas. These are often made by filament winding carbon or Kevlar onto aluminium and sometimes stainless steel liners.

6.4.5. Pipelines and chemical plant

Fibre-reinforced plastics are ideally suited to many situations where aggressive fluids have to be handled. They are very corrosion resistant, and may be used for gaseous bromine, chlorine, and carbon monoxide, and when carbon is used, for concentrated or dilute acids and alkalis. They can match the properties of Hastelloy C, and perform better than stainless steels in many cases. They are, however, limited by relatively low operating temperatures, and their susceptibility to attack by some organic liquids.

Being lighter than metals, composite structures are much easier to transport to the site, and install. They usually require less maintenance than metals, since they do not have to be checked so frequently for corrosive damage, and in some instances their low thermal conductivity reduces the amount of insulation required.

Pressure vessels and pipes made from reinforced polymers must be lined, otherwise they can start to leak at stresses which are very much less than the ultimate tensile strength of the composite. This liner can be metal (suitable for the handling of organic fluids) or polymer (generally used for inorganic fluids). They are usually made by filament winding, and chemical plants make a great deal of use of reinforced plastic pipes. Moreover, the supporting structure for the pipe system is often made with glass reinforced plastic.

Figure 6.25. Mixing and holding tanks and associated pipework made with glass-vinyl ester. (Courtesy of Nemato Composites Inc.)

The materials must be as fireproof as possible. Glass reinforced plastics are more fire-resistant than the plastic matrix on its own because the glass will not burn. If greater fire retardancy is required than that imparted by the glass, fire retardants such as halogenated polymers or antimony trioxide are added to the resin. Fig. 6.25 shows mixing and holding tanks. These were made with glass-vinyl ester. The inside surface uses random glass "veil" which has only 10% fibre with the rest polymer. The next layers are woven glass with 27% fibres and finally the load bearing layers on the outside are filament wound with 55-75% glass fibre content.

Carbon fibres, because of their high cost, are only used in chemical plants where their superior corrosion resistance or excellent specific properties are worth the extra cost, for example in industrial centrifuges. Their electrical conductivity may also be used in applications where a vessel requires electric heat; the heating can be produced by passing an electric current through the fibres.

Probably the largest use of reinforced plastics, however, is for storage vessels, where loading is relatively small.

6.4.6. Infrastructure

Significant load bearing use of reinforced plastics in roads and bridges etc. is a recent development. Here again, wood was the first comer, but steel and concrete are the norm. Bridging the Straits of Gibraltar cannot be done with traditional materials, but might be possible using carbon fibre suspension and light weight composite decking. But as the recent reinforced plastic footbridge across the Thames in London has shown once again, load carrying capacity is only one of many factors in the design of high aspect ratio lightweight bridges. They are prone to suffer from vibrations caused by wind and traffic movement which create instability. Nevertheless, pedestrian bridges made of glass-polyester are now quite common. Mobile assault bridges to be carried on tanks have been made with carbon-epoxy. For as little as 5t weight, they can have a span of 15m.

Figure 6.26. Bridging Muddy Run Creek in Glasgow, DE, using glass reinforced vinyl ester. Parapets are the original concrete, refurbished. (Courtesy of University of Delaware.)

The role of reinforced plastics in bridges for heavy vehicle use has been for repair and protection: "retrofitting". For example, carbon-epoxy reinforcing rods have been used to reinforce concrete decks on motorway bridges. Steel was replaced because it was seriously weakened by corrosion. (Although the alkaline environment of the concrete was expected to protect the steel, the salt we use to disperse ice and snow on our roads seeps into the material and, with accompanying water and air, starts the corrosion process. Since the iron oxides occupy more space than the steel, the concrete cracks very severely, and breaks away allowing faster water ingress.) Glass fibres are not used directly in concrete since they are weakened also. However, glass-epoxy bars have been used successfully as dowels holding concrete paving slabs together. They have proven to be much more durable than the steel used hitherto. Moreover glass-vinyl ester sandwich structures with honeycomb, or other expanded cores, are being used for bridge decks. A program of bridge replacement, starting with relatively small bridges such as shown in Fig. 6.26, has recently been initiated.

Glass-Kevlar-epoxy hybrid I beams have been used to replace steel on highway bridges. The design must take account of the low shear modulus of the composite. This retrofitting has become a major business in the USA now that the Interstate Highways are more than 30 years old.

In addition to bridge decks, the bridge supports are repaired. Carbon-epoxy is sometimes used. Special methods and machines have been developed in order to produce high quality laminated strengthening for the columns and other supports. Woven fibre sheets are impregnated with resins, wound around the column and then cured. Another important use is repair after, and protection against future seismic damage. This is presently an important activity in Japan and California.

Old water mains are also repaired using fibreglass.

6.4.7. Medical applications

The economics of the medical use of materials is quite different from that of most other applications. The major costs in medicine are the facilities: hospital, equipment, etc. And the labour: surgeons, anesthetists, and other highly paid specialists, together with nurses and support staff. The amount of material in a medical application is usually quite small, less than 1 kg, so the cost of even the most expensive materials is usually insignificant compared with the other costs. In addition, there is the time and suffering of the individual requiring the device, and expensive composites can often assist speedy recovery.

A new material cannot be accepted immediately for medical use. A device designed for internal (prosthetic) use has to be checked for unfavourable reactions, both short-term and long-term. The same is true for external (orthotic) devices, though the requirements are generally much less stringent.

Fibreglass is rapidly supplanting wood, leather, and steel for braces (or calipers) for arms and legs. The braces are stronger, lighter, more comfortable, and less noticeable. No straps are needed, and improved cuffs, having air holes, and padded with washable foam, can easily be produced.

Fibreglass is also used for artificial legs and arms. However, the advent of carbon has fostered significant design improvements. Fig. 6.27 shows the internal structure of lower limb replacements. This includes a shock absorber (non-composite) at the top, with below it, the main load support made from carbon-epoxy. The curve in it stores energy which can be released appropriately to reduce the effort of walking. The curved carbon-epoxy side spring adds to the stored energy. The foot itself, also carbon-epoxy, is attached using an elastomer, and restores a spring into patient's step.

Figure 6.27. Internal structure of lower limbs. Carbon-epoxy is the main component. (Courtesy Flexfoot.)

An important advantage for reinforced polymers is their easy mouldability, in view of the wide range of shapes and sizes required. The outer casings of the prostheses are made from plaster moulds of the body parts, using woven reinforcement with open, easily shaped weaves. These prostheses can very easily be made to mimic the appearance of human limbs. Carbon may also be used here because it imparts higher stiffness.

The combination of excellent mechanical properties and the very low X-ray absorption characteristics of carbon fibres is providing a unique opportunity for the development of better X-ray analysis and treatment devices. The angiographic technique (which involves the injection of a radio opaque fluid into the bloodstream, for the location of growths and foreign bodies) needs X-ray pictures to be taken in quick succession (up to six per second). The film has to be placed and moved on very quickly, but held precisely. The key to doing this has been the development of carbon-epoxy compression plates. These can be very precisely moulded from prepreg tape, and the plate combines excellent fatigue properties with low X-ray absorption and good elastic properties.

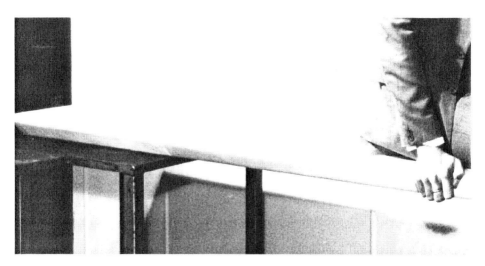

Figure 6.28. Carbon-epoxy support for x-ray therapy. (After Lyons, B.R., and Molyneux, M., 1978, Proc. ICCM2, 1474-92. (Courtesy of the Metallurgical Soc. AIME.)

The computed tomogamy X-ray body scanning technique is used to provide a three-dimensional view of the body. This requires a support which provides great rigidity, so that a person lying on it does not deflect it significantly, and at the same time it must have low and consistent X-ray absorption for radiation passing in any direction in a plane. A carbon-epoxy foam sandwich structure has been developed for this which is highly successful, allowing X-ray resolutions that were not previously attainable. A less exacting requirement is for therapeutic X-ray table tops. These are also made with carbon-epoxy foam sandwiches, and one is shown in Fig. 6.28.

In these X-ray applications, no other material has been found which comes anywhere near the excellence of carbon-polymers.

6.4.8. Sports equipment

As in the case of medical applications, materials cost is not always a major factor in sports. Boron-epoxy has been used for special golf clubs, for example, and carbon-epoxy has become commonplace in racquets for tennis, badminton, squash, etc. In fact, the advent of better materials has had a profound effect on the sports themselves. In men's tennis championships nearly all is decided on $50ms^{-1}$ (110mph) serves, thanks to the efficiency with which modern carbon-epoxy racquets can transfer energy from the arm to the ball. Pole vaulters now reach heights about 25% greater than previously attained thanks to the use of glass-epoxy poles. The key here is the glass fibres, which, with a strength of 2GPa or so, and a breaking strain of nearly 3%, can store large amounts of energy in small volumes with low weights.

Light weight is very important for many sporting applications. Modern racing bicycles are made with carbon-epoxy for this reason, and so are modern racing cars and sculling shells. Also important is the air drag, and using reinforced plastics gives the necessary smooth surface and aerodynamic design. This is supremely important in bobsleds, Fig. 6.29. Moreover, using a tough resin makes the sled more durable, so that it can withstand the bumps and the "monocoque" construction (effectively a single piece enclosure, without internal protrusions) provides better protection for the riders.

Figure 6.29. Bobsled. Aerodynamic design and surface smoothness are overriding considerations which dictate the use of fibre reinforced polymers. Manufacturers claim a 40% reduction in drag compared with metal. (Courtesy of Dow Chemical.)

Fibreglass is still the most widely used composite. Skis usually contain fibreglass, and so do ski poles, bows for archery, gun stocks and butts, hockey sticks, and helmets and face masks. These applications all represent advances over traditional materials due to better durability and improved specific mechanical properties. Again, for the highest performance, carbon is used, and Fig. 6.30 shows some examples.

6.4.9. Other uses

Fibreglass and other reinforced polymers seem destined eventually to take over nearly all uses of metal which do not require the unique properties of the more useful metals, i.e. ductility, hardness, high-temperature resistance, and conductivity. Unreinforced plastics have already supplanted metals in many applications where the low modulus and strength of the polymer are not a disadvantage. (A good reason for the use of polymers in our increasingly energy-conscious era is the low energy required to produce and process them, as compared with ferrous alloys and aluminium, the two most commonly used metals.) Now that reinforced polymers of good quality can be produced reliably and cheaply, these can be expected to take over in areas where moderate to high modulus and strength are required. In addition, the unique properties of some composites (for example the negligible thermal expansion of carbon reinforced polymers) can sometimes increase their advantages over metals at moderate temperatures.

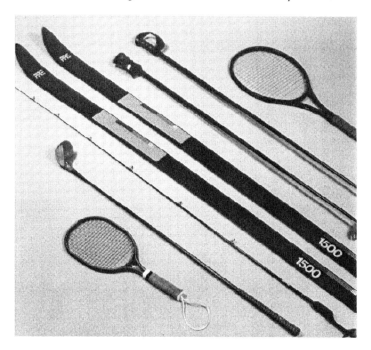

Figure 6.30. Carbon-epoxy sports equipment. (Courtesy of Hexcel.)

Glass fibre reinforced plastics are widely used in agriculture, home appliances, business machines, electrical and electronic hardware, and in materials handling, as well as in the areas already described in more detail above. Fig. 6.31 shows a chair in which braided sleeving was slipped over the core and pulled down tightly to conform to the chair's contour. This made for a very strong, stiff, and durable structure.

Carbon fibre reinforced plastics, being a much more recent development, are much less widely used at present. A use which takes advantage of the negligible thermal expansion of this material is a radar reflector dish used as part of an antenna. This is a sandwich structure, with an aluminium honeycomb core. The thermal stability of the carbon-polymer ensures that the antenna remains accurately tuned to the signal over a wide range of outdoor temperatures.

Another use of carbon fibres that takes advantage of their unique properties, this time light weight, is for moving parts in textile machinery. New machine designs using carbon pultruded bars (which can be made with greater precision than metal constructions) have increased weaving speeds by a factor of ten to twenty compared with conventional machines. The machines also contain injection moulded carbon fibre reinforced nylon. These mouldings can be made very accurately, allowing very precise alignments of the parts during machine operation. In another textile processing machine an injection moulded nylon traverse guide is used. It is shown in Fig. 6.32. The speed of operation of this machine is limited by friction and wear of this part, and the good friction and wear characteristics of the carbon-nylon are a great advantage. In addition, the carbon conducts the frictional heat away from the sliding face more efficiently than the unreinforced polyamide used previously. These advantages are also exploited in carbon-polytetrafluoroethylene (PTFE) bearings used in other machines.

Figure 6.31. Chair stiffened and strengthened using braided sleeving followed by resin impregnation. (Courtesy of A & P Technology.)

Figure 6.32 Carbon-nylon traverse guide for textile processing. (E.M. Trewin, 1978, Proc. ICCM2, 1474 -92. [Courtesy of the Metallurgical Soc. AIME].)

Carbon-epoxy has also proven to be superior to stainless steel in industrial drive shafting, due to good corrosion resistance and higher E/ρ. They, together with glass reinforced epoxies are now widely used in marine propulsion, paper mill drives, large fans and pumps.

Fibreglass is still used for radomes and some of these are very large nowadays: the frontispiece shows the CN Tower in Toronto, which was built primarily as a communications tower and is the first free-standing tower to exceed 550m in height. It houses antennae for numerous microwave links, five television channels at >1Mw and five FM channels at 40kw each.

Figure 6.33 Glass-polyester radome atop Toronto's 550m (1804 feet) Communication Tower. (Courtesy CN Tower Ltd, Canada.)

The radio and TV antennae are housed inside fibreglass tubes up to 7.5m in diameter with 4cm wall thickness and 105m high, shown in Fig. 6.33. Apart from its good microwave propagation properties, the material affords excellent protection from the weather at this height. This can sometimes be very severe. It is built to withstand winds of $400km/hr^{-1}$ and was undamaged by $190km/hr^{-1}$ winds in February 1978 which did a great deal of damage to buildings in Toronto. Another very important property is that ice and snow do not stick to this material. There is thus no danger of it icing up and the built-up ice blocks subsequently falling more than 300 m to earth, and endangering passers-by.

Although the future for reinforced polymers is assured, reinforced metals and ceramics have not yet matured. This is almost certainly only a matter of time. Technical problems, unlike human ones, always seem to be solvable.

Further reading

Hazen, J. (Ed), *High Performance Composites*, published every two months. (Ray Publishing, Wheat Ridge, CO.)
Hazen, J. (Ed), *Composites Technology*, published every two months. (Ray Publishing, Wheat Ridge, CO.)

Problems

You are recommended to solve the following problems in the order given. Data needed will be found in tables in Chapter 2 and other chapters of the book. Where a range of values is given, take the mean value.

(Note: Some of these problems, at least in the same outline form, have already appeared LBFC1 and LBFC2.)

6.1.   Compare the minimum radii that can be used in the weaving of stiff carbon, SCS6, and E-glass fibres with a diameter of 12μm. If the glass fibres are spun, what is the minimum number of fibres that should be used in the yarn if it is to be used for weaving with the maximum possible flexure? Assume that the compressive and tensile strengths of these fibres are the same. For the second part of the question, assume that the spun glass fibres can adjust themselves within the yarn so that the only stresses they experience are due to flexure arising from the curvature of the individual fibres as they are woven around other fibre yarns. (Hint: you also need to make assumptions about the fibre packing in the yarn, and the diameter of the yarn for the tightest weave.)

6.2.   Some stiff carbon fibres, 7.8μm in diameter, were added to S-glass fibres to increase the modulus of the yarn prior to weaving. The yarn contained 204 fibres each with 10.4μm diameter. Calculate how many carbon fibres need to be added in order to marginally increase the yarn load carrying capacity.

6.3.   Use the beam equations in q. 5.2 to show that, for maximum flexural stiffness with minimum weight, $E^{1/3}/\rho$ should be as large as possible. (Note that a high value of EI denotes great flexural rigidity.) Then go on to compare steel, aluminium, and the carbon-epoxy quasi isotropic laminate of q. 3.13 on that basis. (Assume that the laminate density is $1.68 \mathrm{Mgm^{-3}}$.)

6.4.   A business man flies regularly on a 2500km non-stop journey at a height of 900m over the sea. The plane consumes energy at a rate of $95 \mathrm{Js^{-1}}$ $\mathrm{kg^{-1}}$ to keep it aloft and moving at its cruising speed of $210 \mathrm{kmh^{-1}}$. By calculating the fuel consumption, determine the fuel saving when substituting the carbon-epoxy of the last question for aluminium for the wings and fuselage. When made of aluminium the volume of material needed is $0.051 \mathrm{m^3}$. More carbon-epoxy is needed to preserve the flexural rigidity of the structure. The other parts weigh 230kg, and the business man weighs 67kg. The heat of combustion of the kerosene fuel is $52.8 \mathrm{GJm^{-3}}$; its weight may be neglected. Calculate the amount of fuel used in each case, neglecting the work needed to climb up to cruising altitude.

6.5.   Develop a criterion for a material to be used for floors in passenger carrying jet aircraft. The aircraft can make journeys totalling 10 million km. before the floor needs replacing. The floor needs flexural rigidity, and any weight saved can be used to carry extra passengers, giving an increase in income of $0.05 per passenger km. The average weight of a passenger plus baggage and seat is 90kg. Compare an aluminium alloy costing $2kg^{-1} with carbon-epoxy costing $50kg^{-1}, and having the same properties as in the two previous questions.

6.6.   Show that the elastic energy stored in a bar in tension is ½ stress x strain x volume. Hence, determine whether a better material than rubber can be used to drive model aircraft. For simplicity, compare each in tension, and neglect the weight of the system needed to convert linear movement to rotation. Consider silica-, SiC-, and carbon-epoxy instead of natural rubber with a strength of 32MPa, a modulus of 18MPa and a density of $1.13 \mathrm{Mgm^{-3}}$; V_f cannot exceed 0.75 without loss of properties. The properties of the silica fibres are: $\sigma_{fu} = 5.8 \mathrm{GPa}$, $E_f = 72 \mathrm{GPa}$ and $\rho_f = 2.2 \mathrm{Mgm^{-3}}$. Assume $\rho_m = 1.21 \mathrm{Mgm^{-3}}$.

6.7.   Compressed gas is being considered as an energy store. Show that the energy stored is PdV for pressure P and change in volume dV, and hence that, with a perfect gas (for which PV is constant) the energy stored in compressing it from pressure, P_1 to P_2 is $P_2 V_2 \ln(P_2/P_1)$ where V_2 is the volume of the container. Suppose that a gas turbine can be designed to work at 90% efficiency over a pressure range from full pressure P_2 to ½P_2. (At lower pressures its efficiency becomes negligible.) It will work at a remote site, so the gas storage vessel must be as light as possible. The vessel will be a long buried tube, so that the stress in the tube wall is $P_2 D/[2t]$, and uniaxial, where D = tube diameter and t = wall thickness $(t \ll D)$. Derive a figure of merit for the material, and compare the performance of the composites considered in q. 6.6 with the aluminium alloy in Table 2.1.

6.8.   A rotating long thin tube is under stress due to centrifugal forces. For a peripheral rotation velocity of v, and a mass M_T per unit tube external surface area, the forces exert an equivalent pressure equal to $2M_T v^2/D$ where D is the tube diameter. Calculate the stress in the tube, and show that the kinetic energy stored in the tube (half the tube mass

M_T, multiplied by the square of its velocity) cannot exceed the product of half its ultimate tensile strength and the volume of the material in the tube. Now flywheels have been considered for energy storage in moving vehicles. Here the volume of the ring material has to be kept to a minimum. Compare the materials in the two previous questions on the basis of maximum energy stored per unit volume of material.

6.9.   Compare the efficiency of the various forms of energy storage, on a volume basis in questions 6.6, 6.7 and 6.8 with that of diesel fuel, with a potential stored energy of about $60GJm^{-3}$. (This is important for road transport.) In your answer list the various methods in ascending order of effectiveness, giving the energy stored in GJm^{-3} for the most effective in each category. The pressure vessel has a maximum practical pressure of 100 atmospheres, and the volume of the material used to make the container must not exceed one tenth of the volume of the contents.

6.10.   The Ford Motor Company was making a car with a carbon-polymer body, in order to reduce petrol consumption. The carbon-polymer has a density of $1.65Mgm^{-3}$. The body uses $0.035m^3$ of material. The other parts of the car weighed 420kg. How much petrol would be used and what percentage saving could be made, compared with a steel body, using the same volume of material, in an average daily journey to work. In this average journey there is a driver weighing 72kg, no luggage, and no passengers. The distance travelled is 10km, and the journey takes 25 minutes, of which 30% is spent accelerating, 30% decelerating, 20% waiting at traffic lights, and 20% travelling at constant speed. Assume that the acceleration is always at a rate of $1.6ms^{-2}$ up to a speed of $50kmh^{-1}$. The engine is 20% efficient when accelerating, but requires an energy input of $20kJs^{-1}$ to keep it turning over when idling and $95kJs^{-1}$ for travelling at $50kmh^{-1}$. The heat of combustion of petrol is $48.5GJm^{-3}$.

6.11.   A reasonably good cross country skier can work at a rate of 230 watts for long periods. On the level he travels 3.1m with each leg movement. If we assume that all the work goes into accelerating the ski and the leg below the knee from rest to twice the skier's average velocity, calculate and skier's speed on the level with skis made (a) from wood with $\sigma_{1u} = 34MPa$ and $\rho = 0.86Mgm^{-3}$ and (b) strong graphite-epoxy with $\rho = 1.36$ Mgm^{-3}.The skier weighs 70kg, and each foot and leg below the knee weighs 2.7kg. The ski is 2.30m long, and for simplicity we will assume constant, oblong section, with a width, b, of 54mm and a thickness determined by the material strength. Thus it should just support the skier when only the two ends are in contact with the ground.

6.12. A support used with X-ray therapy is fixed at one end and must deflect as little as possible when a patient is on it. It must also absorb X-rays as little as possible. The bed is made as a sandwich, which can be as thick as needed to keep flexure to a minimum. The filling resists shear but does not contribute to the bending moment. Its X-ray absorption can be neglected. The surface skins take all the tensile and compressive stresses. (Consequently the stresses in them can be calculated by equating the external moments with the internal moment coming entirely from the skins). Use a Rule of Mixtures expression for the calculation of X-ray absorption, using the atomic numbers as a measure of the absorption per unit thickness. The stresses in the skins cannot exceed half the ultimate tensile strength. Derive a criterion of excellence, and determine the relative positions of silica-, SiC-, and carbon-epoxy with $V_f = 0.75$, and aluminium. The atomic numbers are: boron 5, carbon 6, aluminium 13. For silica use 10 (the average for Si and two O's, and do likewise for SiC) and for epoxy use 3.6 (an average for hydrogen, oxygen and carbon).

Selected Answers

6.1.   Carbon; 0.41mm: SCS6; 7.35mm: E-glass; 0.13mm:
number of fibres in the yarn; 530.

6.3. $E^{1/3} / \rho$ values: steel; 0.76: aluminium; 1.53: laminate; 2.53

6.5.   Benefit factor = $1909 per kg

6.7.   Merit ratings: silica; 2.28: SiC; 1.08: carbon 1.73: aluminium alloy; 0.11

6.9.   In ascending order of excellence: silica-epoxy tensile bar; $0.179 GJm^{-3}$: silica-epoxy pressure vessel; $1.02 GJm^{-3}$: silica-epoxy centrifuge; $2.22 GJm^{-3}$: diesel $60 GJm^{-3}$.

6.11   Wood; $3.8 ms^{-1}$: carbon-epoxy $4.7 ms^{-1}$.

APPENDIX A   Symbols used in text

Where possible, standard nomenclature has been used. Duplication could not be avoided entirely, but each symbol is described fully in the appropriate part of the text.

A laminate extensional stiffnesses; has subscripts indicating directions
C stiffness; always has subscripts indicating directions e.g. C_{16}, C_{ij}
D constant; diffusion constant;
E Young's modulus; usually has subscripts indicating directions or materials, e.g. E_1, E_x, E_f, E_m, or shorthand for expressions, eg., E_λ, E_ν,
$\boldsymbol{e}_a$ activation energy (Chapter 5)
F force
G shear modulus; usually has subscripts indicating directions or materials
$\mathcal{G}$ work of fracture; usually has subscripts indicating mechanisms and directions
I moment of area
K bulk modulus
$\mathcal{K}$ fracture toughness; usually has subscripts indicating mode or directions etc.
L length
M moment of forces; proportion of water absorbed
N forces per unit length; has subscripts indicating directions
P pressure
P force
Q reduced stiffness; subscripts indicate directions
$\overline{Q}$ transformed reduced stiffness; subscripts indicate directions
$\hat{Q}$ laminate transformed stiffness; subscripts indicate directions
R radius of curvature; distance from centre; fatigue stress ratio
$\boldsymbol{R}$ gas constant (Chapter 5)
S compliance; subscripts indicate directions
$\overline{S}$ lamina transformed compliance; subscripts indicate directions
$\hat{S}$ laminate transformed compliance; subscripts indicate directions
T temperature
T T_g, glass transition temperate; T_m, melting temperature
V volume fraction; subscript indicates material
W width

a crack length
b width
d thickness or fibre diameter
e volume expansion
h length, or height
k constant (has numerical subscripts)
l length
r radial axis; radius
r_c crack tip radius; notch radius
t thickness
u displacement; subscript indicates direction or material
v displacement in the y direction
w displacement in the z direction.
x, y, z directions

Δ	small difference, e. g.
α	thermal expansion coefficient; subscripts indicate material; also constant
β	constant; slope
γ	shear strain; subscripts indicate shear plane or material
ε	axial strain; subscripts indicate directions or material, or limit
η	mutual influence coefficients; $\eta_{xy} = \overline{S}_{16} E_x$, $\eta_{xy} = \overline{S}_{26} E_y$
v	Poisson's ratio; subscripts indicate directions or material
ρ	density
σ	tensile stress; subscripts indicate direction, material, or limit
τ	shear stress; subscripts indicate directions material, or limit
ϕ	angle of axis rotation
ψ	surface energy

Subscripts

0	number, usually indicating an equilibrium or initial value
I	(Roman numeral) indicating crack opening mode
II, III	(Roman numerals) indicating crack shear modes
1, 2, 3, 4, 5, 6	direction or number

A	M_A is the mass per unit area
T	M_T is the total mass
W	R_w is bending radius

a	adhesive
c	compression (e.g. σ_{cu}); also for cracks (r_c), and critical values e.g. σ_c = fracture stress
f	fibre; and in T_f, final temperature
g	for glass transition temperature
i	index
j	joint; also index
$\mathcal{K}$	for conversion to toughness
m	matrix
m	for melting temperature
max	maximum
min, min	minimum
p	polymer
r	radial
t	tensile
tip	tip (crack tip stress)
u	ultimate (for stress or strain)
v	voids
w	water
x	direction
y	direction (single subscript) yield value (second subscript)
$\mathbf{y}$	yield value
z	direction

λ	for thermal shrinkage stiffness
v	constant for thermal shrinkage stiffness constant
∞	σ_x: used to indicate macrostress or stress at a long distance from the crack

Appendix B   SI Units

SI means Systeme Internationale d'Unités, and is an internationally agreed set of units. It is rapidly gaining world-wide acceptance, is replacing local systems such as the British Imperial one, and its North American cousin, and is usually called the metric system.

It is a modified form of the MKS system (Metre-Kilogram-Second) which was introduced in the late 1940's as an alternative to the cgs system (centimetre-gram-second) which had the disadvantage that electromagnetic and electrostatic secondary units (for example for potential difference and current) were many orders of magnitude different, and were inconsistent with the practical volts and amperes used everyday. (The MKS system did not have that disadvantage, nor does SI). It is fully described in ISO 1000.

In SI the basic units include metre (m), kilogram (kg), second (s), and temperature in Kelvins (K). Temperature can also be expressed in Celsius ($^{\circ}$C) To convert K to $^{\circ}$C subtract 273.2.

SI makes a clear distinction between mass and force. For force there is a secondary unit (i.e. one that can be expressed in terms of the primary units). It is called the Newton (N) and is the force required to accelerate 1kg at rate of $1ms^{-2}$. Its units are therefore $kgms^{-2}$. The earth's gravitational field exerts a force of 9.81N on a mass of 1kg at the earth's surface. (The force on the kilogram at the surface of the moon is about 2N).

Table C1. How orders of magnitude are expressed in SI

Power of ten	Name	Symbol
-18	atto	a
-15	femto	f
-12	pico	p
-9	nano	n
-6	micro	μ
-3	milli	m
3	kilo	k
6	mega	M
9	giga	G
12	tera	T

Powers of ten should not be expressed explicitly. Instead a symbol is used. These are listed in Table C1. Sometimes c is used for 10^{-2}, as in cm, but this should be avoided wherever possible. Low speeds, for example, are given in mms^{-1} (millimetres per second), higher speeds as ms^{-1} (metres per second) and very high speeds as kms^{-1} or Mms^{-1} (kilometres per second or megametres per second).

When writing the units, slashes are not used for inverse units. Instead a negative power is used, e.g. ms^{-1} is used for speed, rather than m/s. The axes of graphs should never have statements like: speed; ms^{-1} x 10^5. This is confusing. Use either kms^{-1} or Mms^{-1}, whichever gives the fewer zeros. Table C2 lists some useful conversion factors. To make the meaning of the conversions clear, slashes separate the Imperial and SI units, and orders of magnitude are expressed as powers of 10.

Table C2. Conversion factors between SI and Imperial Units.

To Make the Conversion Multiply by the Appropriate Factor. The Factors are Accurate to at Least One Part in 1000.

Quantity	Multiplication Factor			
	SI to Imperial		**Imperial to SI**	
Length	3.281	ft/m	0.3048	m/ft
	39.37	in/m	0.02540	m/in
Area	10.76	ft^2/m^2	0.09290	m^2/ft^2
	1550	in^2/m^2	0.6452×10^{-3}	m^2/in^2
Volume	35.31	ft^3/m^3	0.02832	m^3/ft^3
	61.02×10^3	in^3/m^3	16.39×10^6	m^3/in^3
Mass	0.9842	ton/tonne	1.016	tonne/ton
	2.205	lb/kg	0.4536	kg/lb
	35.27	oz/kg	0.2835	kg/oz
Density	62.42	$lb/ft^{-3}/Mgm^{-3}$	0.01602	$Mgm^{-3}/lb/ft^{-3}$
Force	0.2248	lbf/N	4.448	N/lbf
Stress and Pressure	0.1450	kpsi/MPa	6.895	MPa/kpsi
	0.06473	tons si/MPa	15.45	MPa/tons si
Fracture toughness	0.910	ksi√in/MPa√m	1.099	MPa√m/ksi√in
Thermal expansion	0.5556	$°F^{-1}/K^{-1}$	1.800	$K^{-1}/°F^{-1}$

In this text stresses and pressures are always given in Pascals (Pa). This is another secondary unit; $1Pa = 1Nm^{-2}$. It is a very small unit indeed; atmospheric pressure is about 0.1MPa. Thus, we have normally to use MPa or GPa. Expansion coefficients are given in MK^{-1} (i.e. micro metres of expansion, per metre original length, per degree Kelvin). Densities are given in the unfamiliar units of Mgm^{-3}, but since $1 \ Mgm^{-3} = 1 \ gcm^{-3}$ this should present few problems.

Appendix C   Some Acronyms

The number of acronyms you can meet in the polymer and composite business are many. Here is a guide. Some of these abbreviations are used in this book when otherwise there would be needless repetition of cumbersome expressions such as coefficient of thermal expansion.

ASTM	American Society for Testing and Materials
BMC	bulk moulding compound
BMI	polyimide (bismaleimide)
CBS	curved beam strength
CFRP	carbon fibre reinforced plastics
CLC	combined loading compression
CTE	coefficient of thermal expansion
CV	coefficient of variation
FEA	finite element analysis
GRP	glass (fibre) reinforced plastics
HDPE	high density polyethylene
KFRP	Kevlar fibre reinforced plastics
LBFC1	*Load Bearing Fibre Composites*, M. R. Piggott, Pergamon, 1980
LBFC2	*Load Bearing Fibre Composites* 2nd Ed^n., M. R. Piggott, Kluwer, 2002
LDPE	Low density poyethylene
MMC	regrettably used for metal matrix composites, including particulate filled
PA	polyamides (nylons)
PAN	polyacrylonitrile
PC	polycarbonate
PE	polyethylene
PEEK	polyetherether ketone
PEI	polyether imide
PES	polyether sulphone
PMC	polymer matrix composites (does not include "filled polymers")
PMMA	polymethyl methacrylate
PPO	polyphenylene oxide
PS	polystyrene
PTFE	polytetrafluoroethylene
PVC	polyvinyl chloride
PU	polyurethene

RIM	reaction injection moulding
RRIM	reinforced reaction injection moulding
RTM	resin transfer moulding
SAN	styrene acrylonitrile
SEM	Scanning electron microscope
SMC	sheet moulding compound
S-N curves	stress amplitude vs log (number of cycles to failure)
SRMM	structural RRIM
TMC	thick moulding compound
UHMWPE	ultra high molecular weight polyethylene

INDEX

Bold numerals indicate the beginnings of sections mainly devoted to the subject.

******

About the Author

The professor grappling with yet another research proposal, August 1997.

Michael Piggott is Professor Emeritus of Chemical Engineering and Applied Chemistry at the University of Toronto. Brought up in Britain, he obtained his BSc in physics, along with an ARCS, at Imperial College, London, in 1951. He then went on to graduate studies of thin film chemical reactions, and friction and wear of surface films, for his PhD, again at I.C. After a spell in industrial research and production in England he went to work in Canada at the Chalk River Nuclear Laboratories (Atomic Energy of Canada Ltd. [AECL]) in 1964.

There he was put to work in the new area of composite materials. AECL had great hopes for reinforced aluminiums for pressure tubes for nuclear reactors. So he applied his thin film knowledge to the potentially devastating reaction between silica, the only fibre worthy of consideration for nuclear reactors at that time, and aluminium, the most hopeful matrix material for use in the hot and radiation-filled nuclear reactor environment. He made the discovery that the reaction was immediate, thus conflicting with prior work. He thus set off on the lonely road of enquiry followed by those without preconceptions. (Unlike his predecessors, he merely did the experiment with pure materials, and with no impediments, e.g. oxide films.)

He re-examined the theory of composites and managed to publish enough papers to be appointed Associate Professor in the new Materials Centre at University of Toronto in 1968. Arriving there, he was surprised to find that he was in Chemical Engineering. A blessing in disguise. He was encouraged to learn about polymers, and realised their great potential for reinforcement. So he gradually switched over from reinforced metals. Plastics are so much easier.

He proceeded to amass papers to ensure that he was on an ascending spiral of more research papers ⇨ more research funds ⇨ more graduate students ⇨ more experimental results ⇨ more papers ⇨ more research funds .... ever upward, till the granting agencies decided that he was in his dotage at the advent of the new millenium.

Now he has enough papers, books and citations to appear in Marquis Who'sWho in the World, along with other notorious people like Colonel Gaddaffi and Fidel Castro. Perhaps justly notorious: he hasn't endeared himself to the Oracles of Composites, A. Kelly and S.W. Tsai. He has shown that the former led researchers down a blind alley in fibre/matrix interface studies and the latter gave angle ply laminates such a bad name that they haven't ever been used in the aircraft industry, despite their irrefutably sterling qualities.

For relaxation he has climbed six virgin peaks in the Canadian Arctic, built two cottages, has been editor-in-chief of the North Renfrew Times, sung with the Canadian Opera, and enjoys the occasional company of his six far-flung grandsons (no granddaughters, alas).

The author feeling on top of the world in the Dauphiné region of the French Alps.
Summer ~ 1952.

Michael R. Piggott, November 2006